Memory Palace

Learn the Secrets to Build Memory Palace and Finally

(Brain Training Guide to Memory Improvement, Essential Study Techniques)

Robert Hernandez

Published By **John Kembrey**

Robert Hernandez

Memory Palace: Learn the Secrets to Build Memory Palace and Finally (Brain Training Guide to Memory Improvement, Essential Study Techniques)

ISBN   978-1-998927-81-4

No part of this guidebook shall be reproduced in any form without permission in writing from the publisher except in the case of brief quotations embodied in critical articles or reviews.

Legal & Disclaimer

The information contained in this book is not designed to replace or take the place of any form of medicine or professional medical advice. The information in this book has been provided for educational & entertainment purposes only.

The information contained in this book has been compiled from sources deemed reliable, and it is accurate to the best of the Author's knowledge; however, the Author cannot guarantee its accuracy and validity and cannot be held liable for any errors or omissions. Changes are periodically made to this book. You must consult your doctor or get professional medical advice before using any of the suggested remedies, techniques, or information in this book.

Table Of Contents

Chapter 1: Why Does The Memory Palace Technique Work?

Contrary to well-known perception, this is not a modern-day-day method the least bit – even ancient Romans might also practices it. It is the famous method utilized by the pinnacle champions inside the reminiscence fields – this approach is what made it possible for Dominic O'Brien to memorize a series of 2808 gambling playing cards after having taken into consideration every card once best.

Hollywood has additionally picked up on this with TV's, The Mentalist, Patrick Jayne, being really unstoppable using this technique.

Okay, now allow's deal with the fallout from those to statements – superb, The Mentalist is a fictional series (a top notch one but fictional) and Dominic O'Brien has dedicated years of his existence to honing his memory capabilities. Neither one is your standard example or perhaps a few element together with you, proper?

To an extent, that is real but the clean fact is that everyone may also additionally have the ones top notch powers of don't forget certainly thru getting to know the reminiscence palace technique. You virtually are brilliant restrained via your very very own preconceived notions approximately what can and can not be finished.

Think about this for some seconds – inside the first 5 years of life, we all wished to investigate to walk, talk, interact with each extraordinary, play, and so forth. We have been digital sponges and could soak up big quantities of statistics. And every unmarried one parents turn out to be as soon as a infant with the potential to take in massive portions of records over a completely brief duration.

So, what went incorrect? Do we get greater silly as we emerge as old? That honestly does no longer make any revel in – I apprehend that I, for one, am smarter than a five 12 months vintage. The reality is that we do now not ever lose the capacity to absorb those large quantities of information. What does get up is that we not want to and, as a quit result, get a hint lazy or fall off form. Even regardless of the reality that we do get out of the dependancy of the use of this potential, we do soak up extensive portions of facts every day with out truely noticing it. It is simply that after we're beneath a bit of pressure to don't forget the data, we essentially can't discover it in among all the different stuff floating spherical in our brains. And that is extremely good information – it way that we've the ability, if we take a bit time to discover ways to use it again.

Let's cross another time to me, for example, I even have turn out to be going on a revel in to Dubai and decided that I wanted to study to speak Arabic as a stop end end result. I failed miserably – reading a couple of phrases like "Thank you" and "Greetings. No depend how regularly I repeated the phrases, the statistics may also need to first-rate stick in my thoughts for a couple of minutes at most. Eventually I gave it up as a out of place reason.

Then I discovered out greater approximately the reminiscence palace technique and commenced to learn how to use it. I decided that I have to try to examine some other language the use of this method and I found that studying this time round have become lots less complex. I actually have due to the truth began out mastering to speak Mandarin. (Slowly of route, it's far a hard language to have a look at.) For someone who only some years inside the beyond couldn't even recollect what she'd eaten for breakfast some days earlier, this become a notable fulfillment.

My issue in regarding this is that if I can do it, I apprehend that you can as well. I simply have a totally energetic mind and one that can be effortlessly distracted – part of the motive that I used to have hundreds hassle remembering subjects. With the memory palace approach, I even have an anchor in my thoughts for critical matters that I want to don't forget and a way to get right of entry to them as and even as vital.

I actually have additionally taught everybody I apprehend how the approach works and the consequences have be pretty outstanding. My exceptional friend's daughters are each in high college and can not believe how easy it is to recollect topics using this method – they noticed a marked growth of their grades, virtually due to the truth they had been no longer having to go through subjects again and again another time.

And it isn't best in subjects like statistics that the girls are excelling – math and era additionally require the memorization of masses of unique requirements and equations and the reminiscence palace is fantastic for those moreover.

Why It Works

The memory palace approach works due to the truth it's miles based mostly on the usage of our spatial reminiscence – as human beings, we are very good at remembering places that we understand. Building a acquainted shape lays the concept to maintain the facts in makes it plenty much less complex to don't forget that facts in a while.

You can use the indoors of your house, the path that you stress to art work on a each day basis - as long as it is acquainted to you, this method will artwork.

Your thoughts is bodily created to help you interact with the outside global – that will help you to hold information critical to survival. How many purchasing lists or exam cramming periods did your common caveman want, for example?

Take benefit of this need for spatial stimulation and you may do thoroughly on the subject of remembering what you need to due to the fact now you have got were given a way to without a doubt feed it into your mind in a manner that makes experience to it.

You can select to group certain sorts of memories collectively and assign a place for them for your mental palace in region of the same antique jumble of records this is certainly filled in without regard for having to keep in mind it later. You will have a look at the reminiscences which you need to get admission to and additionally at memories that may be associated.

Chapter 2: What is a Memory Palace?

A reminiscence palace is an area that you visit to your mind, an area that you can visualize on foot thru, with numerous little nooks and crannies where statistics may be saved. The concept is that you'll be capable of take into account the records that you want really by manner of the use of visualizing the right section for your memory palace. For a few humans, this can be a unmarried room, for others, their entire town.

Don't get too stuck up inside the imagery itself – there aren't any right and wrong answers right proper right here. The best actual rule is that it should be somewhere that you understand very well.

You can choose out to apply your own home as a reference point or a series of diverse locations. Perhaps, in case you seize the bus to paintings, you can store statistics to coincide with each of the stops alongside the manner. The key's to find a location which you apprehend in reality properly and that you may visualize on your mind. As lengthy as you are able to do this, you have already got the whole thing which you want to have a reminiscence magnate.

And therein lies the beauty of this device – no extra trying to offer you with clever little sayings or songs to help keep in thoughts subjects, no extra repeating statistics till you're bored to tears and no extra forgetting half of what became to your grocery listing because of the fact you forgot to take it purchasing with you.

When you've got got brought the statistics that you need to undergo in thoughts all you could need to do to bear in thoughts it's far to visualize going into the applicable area of the reminiscence palace and pulling it out once more.

I truely need to pressure one extra time that the reminiscence palace isn't always commonly a constructing – use some thing suits you. It is crucial that there are various precise rooms/ stops, and so on. Alongside the way because of the fact you need to categorize the records and region it in the right spot for do not forget later. Let us say, for example, which you choose to use a single room – maybe your bed room – remember it for a minute, there are although a number of locations to keep stuff in for your mattress room aren't there?

Look at storing data in your cupboards and arrogance however furthermore keep in mind less possibly locations like on top of the pallet, within the lower back of the curtains, underneath the lampshade, and lots of others. If you examine it this way, there are a whole lot of places that you could save records and so you can shop an entire lot of records as nicely.

Your mind loves this method – with the useful resource of way of pairing the reminiscences with spatial recollections, you are telling your mind that this facts is simply as crucial as knowing the manner spherical your house or a way to get to artwork is.

Chapter 3: Building Your Own Memory Palace

Start out thru using your bed room or a unmarried room in your private home to start out with – that may be a area which you apprehend well and it may be a available area to begin working towards this method in a smaller place. Once you've got got gotten the draw close of it, you may make the palace bigger – by using using incorporating your whole residence, or each other location that you are extraordinarily acquainted with. You should, if you like, pick a place that you have imagined but I simplest suggest doing this if the imagery can be very top notch on your thoughts. This technique has an inclination to artwork terrific while you artwork with an area which you were to and skilled so you can keep in thoughts the numerous sensory factors of the revel in.

The greater element that you can upload on your memory palace and the more facts approximately the place that you may upload, the greater information you may be able to maintain there as nicely.

Start By Imagining The Place

Now, close your eyes and, on your thoughts's eye, be given as real with the reminiscence palace that you have decided on. What facts can you see? What tremendous subsections are there? If you run into blocks, open your eyes and go searching and refresh your memory as to what in reality is there. Remember as hundreds detail as feasible – factors of hobby, sounds and smells can all help to make the reminiscence absolutely remarkable in your mind. Even keep in mind the feelings associated with the area as a way to make it all of the greater real on your thoughts.

Pick Your Route

If you want to do not forget things in a particular order, you'll set up an actual roadmap in your reminiscence palace in anything manner makes sense to you. You will want to pick out an appropriate route for what it is that you want to don't forget. Then, as you float alongside the route, matters become easy to preserve in thoughts in ideal order.

Even if you do now not need to bear in mind some issue in a specific order, it is though an extraordinary concept to installation a route for travelling via your memory palace. That manner, it does truly become masses a good deal less tough when you want to do not forget some component.

Basically in terms of recalling topics, you may usually stroll alongside this course until you come back to the segment within the palace that shops the information which you need. If you are in a hurry, you are extra than welcome to bypass without delay to the section that you want.

Where You Will Store Things

You now need to determine out wherein matters are going to be saved inside the memory palace. You may also want to, for instance, determine to put topics pertaining the kitchen inside the kitchen of your reminiscence palace. Find logical connections for top notch kinds of information in the course of your memory palace.

It is also essential to ensure that no person area is so like some different that they will be incorrect for it. You may additionally want to make this step as clean or complex as you want. Maybe you in reality need to consider a small list of information, then make each room in your private home a brilliant place. You can constantly add in more information via looking at the entirety within every room as you cross alongside.

Think of it as some thing similar to the submitting device which you use to your computer – you may have the primary folder (your entire memory palace); sub-folders (in which the information to be stored is divided amongst difficult categories and folders within the ones (locations wherein the more particular records will skip.) As along aspect your laptop, the smaller the file of data, the a amazing deal much less the danger that you need numerous folders and sub-folders.

Ready, Set

We aren't pretty geared up to move sincerely however. You now need to make a drawing of your memory palace and the route that you'll take alongside the manner. Mark out all of the places which you have selected as places to cowl stuff in and stroll the course that you have decided on a couple of instances – taking in as lots detail as you are capable of. Then commit the whole plan to memory.

Now visualize in truth going alongside the course, taking care to fill in as many details as you are capable of as a way to make this a real picture to your mind. Check how you've got finished toward the authentic drawing that you did and, in want, exercise going thru it on your thoughts time and again yet again to make sure which you apprehend every and every single stopping component, and so on.

Practice on the identical time as you're far from the physical place as nicely, without relying on your notes the least bit.

Once you have were given managed to successfully visualize the entire reminiscence palace to your mind, you are prepared to transport right now to absolutely starting to memorize topics.

Chapter 4: Using Your Memory Palace

Okay, now you may get proper all the way down to employer. At this degree, the palace need to be nicely-rehearsed in your thoughts and also you have to be with out issues capable of visualize it. Now it's time to begin assigning the unique records to splendid regions inside the palace.

Working in a logical manner right right here may be of super benefit to you. Let us say, for instance, which you need to memorize a speech, it makes sense that you need to place it inside the right order and in ability segments. Break it up into critical phrases and visually located them in vicinity as you circulate along your memory palace path. Do not be scared to spread out all the manner along the direction and do make sure which you have memorable chunks of the speech spread out inside the proper order.

If you do no longer want to keep in mind topics in any specific order, you may region the objects everywhere within the reminiscence palace. Again, I suggest locating a few logical purpose almost about putting the objects worried. For instance, let us say which you needed to undergo in thoughts a grocery listing with the following on it:

Tile cleaner; oranges, rest room paper, deodorant.

A appropriate way of remembering this listing may be to keep the tile cleaner inside the rest room, the oranges within the kitchen, the relaxation room paper within the rest room and the deodorant inside the bed room.

When you're starting out, it does make experience to group like gadgets collectively as having to go through particular rooms to discover what is stored there can be hard. What I suggest here is that during case you have got turn out to be cleaning assets, as an example, it is better to institution them all in one place, rather than setting them into the man or woman rooms.

As you improvement, this may come to be a whole lot much less and lots less of an trouble and you'll discover that you are capable of positioned the objects anywhere that you like.

On Symbolism

The thoughts is simply as adept at remembering symbols as it is at remembering words and numbers. Whenever possible, cut lower back at the amount of terms used so you have a higher risk of remembering everything that you need to.

Storing the devices in a logical location inside the reminiscence palace is one detail, locating a image that let you to undergo in thoughts them is going to take your take into account up a few notches.

For instance, everybody recognize that the Texas is the Lone Star usa. Perhaps as opposed to putting Texas on your memory palace, you could visualize a film star, all on its very very own. By the identical token, Wisconsin is associated with cheese and California with surfing.

It is lots a lot much less difficult to bear in mind shorter snippets of information and you could frequently effectively shorten the amount of statistics through using a image in place of the actual terms. The mind moreover tends to be masses more visually introduced approximately and so pictures is probably easier to preserve in mind than actual phrases.

You Have Permission To Day Dream

When I modified into at university – a long, long time inside the past – one complaint that came up over and over once more become that I modified into susceptible to day dreaming. Now searching decrease returned I understand that this have become now not a awful problem in any respect but hundreds people had been raised to trust that it changed into. From in recent times, you are allowed to day dream, specifically in terms of things that you need to don't forget.

What you want to go through in thoughts is that the thoughts is without issues bored and is constantly seeking out extra powerful strategies to do topics. Remember the last time which you posted out your New Year's Resolutions and stuck them subsequent to the relaxation room reflect? If you're like most humans, you have been reminded of your desires each time you found the poster for the following week or so. When your mind had been given fed up in the photo, it stopped consciously noting it so no matter the truth that you can had been staring immediately on the photograph, it did not make a whole lot of an impact for your mind.

It works in hundreds the same way when you are trying to look at topics by rote. You repeat the equal phrase over and over another time for your mind till your mind turns into fed up. Whilst you can keep the know-how over the quick-time period, this is not an extended-term answer – take a look at in in step with week or how a whole lot of the facts you have retained through this approach.

Creativity Is Key

If you need to outfox your thoughts, you want to get extra modern almost about memorizing topics. You need to also visualize the devices on the list in as revolutionary a way as viable – as an example, allow us to mention which you want to recollect that the address is 124 Church Lane, you can begin via setting the 124 in its very very own room in your memory palace. You can also want to then keep in mind a spear that seems like the no 1 and bear in mind it piercing the neck of a swan – to symbolize the variety 2 after which believe the swan breaking up into 4 portions.

You can also use mnemonics to make it less difficult to bear in mind records this is going into your reminiscence palace. This entails the usage of the primary letter for each phrase and changing it right into a sentence that you may greater without trouble be capable of take into account. For example, if you need to bear in thoughts the order of the planets in our solar device, you will possibly say, "My very energetic monkey jumps suddenly under Neptune." (Mercury, Venus, Earth, Mars, Jupiter, Saturn, Uranus, Neptune.) Again right right here, the greater absurd the sentence is, the much less tough it's far going to be to consider.

Now which you have stuffed up the rooms of your palace with various photos, symbols, and so on. You want to stroll via this on your mind and make sure which you go through in mind the whole lot in it. As you look at each room, ensure to phrase wherein the entirety is. As you use this technique increasingly regularly, you could find that your thoughts routinely pals the facts that you introduced in in a herbal manner. You will just need to stroll as a great deal because the bed room door on your memory palace to be reminded of what is in there.

Practice walking through the palace some times and recalling the records. Do make certain to comply with the course which you decided to put out and do maintain at it so that you in addition imprint the reminiscences. Eventually, this can grow to be 2nd nature and recalling is probably definitely a do not forget kind of locating the proper location on your private reminiscence palace.

Over time, you can usually assemble more memory palaces. If you're just the use of your palace for remembering lists in the quick time period, that is virtually a rely or getting rid of the statistics that is no longer essential and changing it with the new information.

Alternatively, pick out new routes or assemble a very new memory palace. You also can use precise memory palaces to help you do not forget numerous matters – as an instance, possibly you would really like to have one memory palace for business enterprise contact numbers and one for non-public contact numbers. You can also have as many reminiscence palaces as you want, so long as you take the time to set each up well at the outset.

Your brain is a lot greater great than you may ever apprehend. Take the World Memory Championships, for instance. It is right here that the quality ranked competition correctly control to memorize all the cards of 20 decks, shuffled up in an hour or an lousy lot less. The shorter sports activities can encompass the memorization of about 500 numbers determined on at random, in fifteen mins or a whole lot much less. Pretty wonderful, right?

What is even greater excellent is that most people are able to the right equal feats – the distinction amongst us and those competition isn't always that they'll be masses greater sensible or maybe that their reminiscences are higher. Where the difference is to be had in is in the way that they discover ways to keep information for smooth endure in mind later. They moreover use strategies like mnemonics and reminiscence palaces with a purpose to help with universal don't forget.

The genuinely, simply first-rate records is that you could do the identical – even in case you are currently this sort of folks that would leave their heads behind if it wasn't screwed on, you may moreover improve your maintain in mind capabilities. It will definitely take a hint time, try and staying power to your issue.

Chapter 5: Trouble Shooting and Tricks

Okay, so now you understand what it is which you want to do – the truth of the matter is that it isn't always constantly going to be clean crusing. In this financial disaster, we're capable of undergo some of the matters that would skip wrong and come up with the pointers which you need to get over the ones bumps in the road.

I Just Cannot See It

In this situation, the adage, "If earlier than everything you don't achieve success, try to try yet again" can be very apt. Some humans are going to get this proper their first time spherical but there aren't masses of those humans.

I need to confess that when I first tried the memory palace approach, I couldn't get it proper – the visualization itself become hard for me in the beginning and I discovered out that I am greater without issues able to visualize the general

28

appearance of factors than the actual information. For me, drawing the direction helped a sincere quantity however I did want to exercise pretty a bit so one can certainly get the program going for walks because it must for me.

If you're finding that it's far hard to visualise things first of all, why no longer installation an real excursion of your reminiscence palace – all you'll want is a digital digicam/ cellphone capable of taking video. Walk thru your direction at least as soon as whilst filming after which you could do this digital tour even as the mood takes you. This can be an excellent manner to assist with the visualization manner, in particular if you are scuffling with with it early on.

Alternatively, sluggish down a piece – It is probably tempting to need to use your whole town as a memory palace but the larger the gap for the palace is, the greater difficult it will probably be to visualize, at least initially. Start off with one room or, if that is too much, even just the usage of a corner of the room or a bookcase as a setting out aspect.

Other than that, you do want to practice as hundreds as feasible.

I Need To Memorize New Information

Perhaps the number one palace worked so properly that you have determined which you need to memorize even extra. That is great and you can effects do that via using each adding the information in potential chunks to the rooms that you have or via deciding on a completely precise route altogether.

The path which you choose out out will depend on how important the vintage data changed into and the way the state-of-the-art facts relates to it, if it does at all.

For example, probable you have been reading approximately Ancient Egypt and memorized all of the pharaohs of one specific dynasty. Now that you are getting further in your studies, you have got determined to memorize information approximately their queens and gods as well.

The names of the queens could possibly in form quite nicely with the names of the pharaohs and will be a logical extension – you may store every with the considered necessary pharaoh with little problem. The names of the gods, however, are a greater commonplace topic and can require a room or corner in their non-public.

Perhaps then you definately in reality want to memorize your grocery listing – this has now not anything to do with ancient Egypt and so it'd want to be saved in a brilliant area once more. Alternatively, you can moreover set yourself an entire new memory palace.

I Have A Lot Of Material To Get Through

This is a grievance that I get quite regularly and over again I want to remind you that that little 5-12 months antique which you as soon as have been is somewhere in there. You do have the functionality to get thru mountains of facts. Still now not happy – reflect onconsideration on all of the data that comes at us every unmarried day inside the shape of social media, traditional media, TV, and so forth. We are in no manner definitely unplugged anymore and your brain handles all of that definitely high-quality. You can control huge portions of facts and also you really need to head again to this recognition yourself.

If you are analyzing quite a few information proper away, it's far exceptional to interrupt it up into smaller programs and shop it in this way. Back to the Egypt example – you can, as an example, break up up the pharaohs and their queens into one phase and memorize that, then add inside the gods and goddesses in a separate segment.

Then add within the foremost cities, industries, and so on.

If the whole lot is broken up into smaller bits, you can locate that you are going to find it hundreds less difficult to manipulate.

I Still Cannot Remember Or I Get Mixed Up

There are a few humans who've arise to me and stated things like, "This is complete garbage, it does no longer work." In maximum instances whilst this takes place, it's far because you haven't finished sufficient preparation in terms of your memory palace, or you have not defined the hyperlinks among the records and the area that it's far being stored well.

Start with looking at how you have were given were given placed the information inner your memory palace. Is it dotted everywhere in the vicinity? If you have got have been given similar batches of information at numerous locations interior your palace, it is able to be difficult to don't forget which batches are in which and this could cause confusion.

I may say that spherical about ninety% of memory lapses may be attributed to not paying enough interest and if you preserve having problems with consider on the equal time as the use of this technique, you need to have a test your approach.

Have you in fact visualized the memory palace and brought the time to stroll through it to your mind? Have you positioned the information there in potential chunks and in a logical spot? Have you given yourself sufficient time to nicely absorb the records, without being distracted?

Chapter 6: Some Fun Facts That You Can Memorize And Wow Your Friends With

In this financial disaster I am going to provide you some lists of facts that you could memorize as a exercising run. The lists range from extended to brief and so must come up with a few outstanding exercise. Memorizing the longer lists is likewise a groovy birthday celebration trick – mission your friends to appearance if they could beat you.

Start with the much less complicated bodily games and progress to the greater tough ones as you beautify.

Easy

The Seven Deadly Sins

Lust

Sloth

Gluttony

Greed

Wrath

Envy

Pride

The Seven Virtues

Chastity

Charity

Diligence

Patience

Kindness

Humility

Temperance

The Ten Longest Rivers

Nile

Amazon

Mississippi-Missouri-Red Rock

Yangtze

Ob

Yellow

Yenisei

Parana

Irtish

Zaire

The Seven Wonders Of The World

Great Pyramid of Giza

Hanging Gardens of Babylon

Statue of Zeus at Olympia

Temple of Artemis at Ephesus

Mausoleum of Maussollos at Halicarnassus

Colossus of Rhodes

Lighthouse of Alexandria

The Tallest Mountains inside the World

Everest

K2

Kangchenjunga

Lhotse

Makalu

Cho Oyu

Dhaulagiri

Manaslu

Nanga Parbat

Annapurna

The Ten Biggest Countries

Russia

Canada

United States of America

China

Brazil

Australia

India

Argentina

Kazakhstan

Sudan

Countries With The Highest Population Rates

China

India

United States

Indonesia

Brazil

Pakistan

Nigeria

Russia

Bangladesh

Japan

Intermediate

The Ten Commandments

There can be no exceptional gods but me

You shall no longer worship faux idols

Do no longer take the Lord's call in useless.

The Sabbath is to live holy.

Honor your father and your mom

You shall not kill

You shall now not covet your neighbor's companion

You shall not go through fake witness closer to your neighbor

You shall no longer steal

Be not resentful

More Difficult

American Presidents

George Washington

John Adams

Thomas Jefferson

James Madison

James Monroe

John Quincy Adams

Andrew Jackson

Martin Van Buren

William Henry Harrison

John Taylor

Millard Fillmore

Franklin Pierce

James Buchanan

Abraham Lincoln

Andrew Johnson

Ulysses S. Grant

Rutherford B. Hayes

James Garfield

Chester Arthur

Grover Cleveland

Benjamin Harrison

Grover Cleveland

William McKinley

Theodore Roosevelt

William Howard Taft

Woodrow Wilson

Warren Harding

Calvin Coolidge

Herbert Hoover

Franklin D. Roosevelt

Harry Truman

Dwight Eisenhower

John F. Kennedy

Lyndon Johnson

Richard Nixon

Gerald Ford

Jimmy Carter

Ronald Reagan

George Bush

Bill Clinton

George W. Bush

Barack Obama

Challenging

Best Picture Oscar Winning Movies from 1928 - 2014

1927-28 - Wings

1928-29 - Broadway Melody

1929-30 - All Quiet on the Western Front

1930-31 - Cimarron

1931-32 - Grand Hotel

1932-33 - Cavalcade

1934 - It Happened One Night

1935 - Mutiny on the Bounty

1936 - The Great Ziegfeld

1937 - The Life of Emile Zola

1938 - You Can't Take It With You

1939 - Gone With the Wind

1940 - Rebecca

1941 - How Green Was My Valley

1942 - Mrs Miniver

1943 - Casablanca

1944 - Going My Way

1945 - The Lost Weekend

1946 - The Best Years of Our Lives

1947 - Gentlemen's Agreement

1948 - Hamlet

1949 - All the King's Men

1950 - All About Eve

1951 - An American In Paris

1952 - The Greatest Show on Earth

1953 - From Here to Eternity

1954 - On the Waterfront

1955 – Marty

1956 - Around the World in eighty Days

1957 - The Bridge at the River Kwai

1958 - Gigi

1959 - Ben-Hur

1960 - The Apartment

1961 - West Side Story

1962 - Lawrence of Arabia

1963 - Tom Jones

1964 - My Fair Lady

1965 - The Sound of Music

1966 - A Man for All Seasons

1967 - In the Heat of the Night

1968 - Oliver

1969 - Midnight Cowboy

1970 - Patton

1971 - The French Connection

1972 - The Godfather

1973 - The Sting

1974 - The Godfather Part II

1975 - One Flew Over the Cuckoo's Nest

1976 - Rocky

1977 - Annie Hall

1978 - The Deer Hunter

1979 - Kramer vs Kramer

1980 - Ordinary People

1981 - Chariots of Fire

1982 - Gandhi

1983 - Terms of Endearment

1984 - Amadeus

1985 - Out of Africa

1986 - Platoon

1987 - The Last Emperor

1988 - Rain Man

1989 - Driving Miss Daisy

1990 - Dances With Wolves

1991 - The Silence of the Lambs

1992 - Unforgiven

1993 - Schindler's List

1994 - Forrest Gump

1995 - Braveheart

1996 - The English Patient

1997 - Titanic

1998 - Shakespeare in Love

1999 - American Beauty

2000 - Gladiator

2001 - A Beautiful Mind

2002 - Chicago

2003 - The Lord of the Rings: The Return of the King

2004 - Million Dollar Baby

2005 - Crash

2006 - The Departed

2007 - No Country for Old Men

2008 - Slumdog Millionaire

2009 - The Hurt Locker

2010 - The King's Speech

2011 - The Artist

2012 – Argo

2013 - 12 Years A Slave

2014 – Birdman

Chapter 7: Some Additional Exercises For When You Are More Advanced

Word Soup

Grab a dictionary and open at random – pick the primary phrase that catches your eye and write it down. Do this for a whole of 15 times after which memorize the list in five minutes or heaps less.

To take it up every other notch, furthermore word the meanings of the phrase and memorize those as nicely. You have 15 minutes for this exercising.

Friendly Challenge

Get a pal or neighbor to put in writing down 15 terms that they count on which you can't realize. You have five mins to memorize this listing. Make it more difficult via getting them to increase the considerable fashion of terms that they write down or have them write longer, extra hard terms down.

Memorize them in the order that they've been written in.

Phone Book Challenge

Open the cellular phone ebook to any web web page which you like and memorize all of the numbers in the first column. Make it extra tough with the aid of developing the range of phone numbers which is probably memorized and lowering the amount of time on the manner to do this in.

Take it up some notches by way of using manner of remembering the names and contact numbers. Add the addresses in later if you want.

Card Shark

Shuffle a full deck of gambling gambling cards well and memorize them so as. You have half of an hour for this workout. Make it tougher thru the use of such as in a second and 1/three deck in a while.

Mobile Challenge

Quick test – how a number of the phone numbers on your mobile mobile cellphone do you certainly apprehend? If you needed to lose your cellular telephone tomorrow, how many would you be able to get higher? Now, once I become developing up (returned within the dark a long time earlier than the invention of mobile phones and pace dialing) we virtually had to memorize cellphone numbers or write them down. Back then, I knew what the numbers of all my friends and own family had been. In all honesty, that is not the case and, if I misplaced my cellular cellular telephone now, I may additionally high-quality have the capacity to call my mom because her variety is the most effective one that I do apprehend.

However, if you are searching out a project, why no longer memorize all of the numbers to your cellular cellphone?

Trivial Pursuit

If you have got a Trivial Pursuit recreation languishing behind the cupboard, why no longer pull it out and study a few trivial records? You can memorize complete gambling cards or a set of questions in a single topic – some element you want. At the very least, it is a incredible way to become Trivial Pursuit champion at the subsequent family reunion.

State Of The Nation

US State	State Capital
Alabama	Bernard Law 1st viscount montgomery of alamein
Montana	Helena
Alaska	Juneau
Nebraska	Lincoln
Arizona	Phoenix
Nevada	Carson City
Arkansas	Little Rock
New Hampshire	Concord

California	Sacramento
New Jersey	Trenton
Colorado	Denver
New Mexico	Santa Fe
Connecticut	Hartford
New York	Albany
Delaware	Dover
North Carolina	Raleigh
Florida	Tallahassee
North Dakota	Bismarck
Georgia	Atlanta
Ohio	Columbus
Hawaii	Honolulu
Oklahoma	Oklahoma City
Idaho	Boise
Oregon	Salem
Illinois	Springfield
Pennsylvania	Harrisburg

Indiana	Indianapolis
Rhode Island	Providence
Iowa	Des Moines
South Carolina	Columbia
Kansas	Topeka
South Dakota	Pierre
Kentucky	Frankfort
Tennessee	Nashville
Louisiana	Baton Rouge
Texas	Austin
Maine	Augusta
Utah	Salt Lake City
Maryland	Annapolis
Vermont	Montpelier
Massachusetts	Boston
Virginia	Richmond
Michigan	Lansing
Washington	Olympia

Minnesota	St. Paul
West Virginia	Charleston
Mississippi	Jackson
Wisconsin	Madison
Missouri	Jefferson City
Wyoming	Cheyenne

Chapter 8: More On Mnemonics

Mnemonics are gadgets that resource in preserve in mind. If you're analyzing large quantities of records, combining this form of memory techniques with the reminiscence palace makes experience.

Word Mnemonics

We have already been via the word mnemonics – in which you're taking the letter from every word and create new phrases with those so that you have to make a sentence this is easier to don't forget. Another example of this form of mnemonic tool is: "Please excuse my high-priced Aunt Sally" in an effort to don't forget the order in which to address math functions inner an equation – Parenthesis, Exponents, Multiply, Divide, Add and Subtract.

Music Mnemonics

I recognize that in advance I did say which you didn't should make up any greater silly songs to don't forget things anymore. That said, tune presents a in addition measurement to hold in mind as it is straightforward to don't forget. If you have got were given lengthy lists, placing them to a catchy song can move an prolonged manner to supporting maintain in mind. Think approximately the alphabet music you located even as you have been younger. Have you ever forgotten it? (I apprehend that each time I recite the alphabet in my mind, it is to the tune of the alphabet music.)

Name Mnemonics

This is just like the word mnemonics besides that we form a word/ call with the number one letters of the phrases in vicinity of developing a sentence.

To take into account the colors of the rainbow, as an instance, we want to apply Roy G Biv – Red, orange, yellow, green, blue, indigo and violet.

Rhyming Mnemonics

Because the phrases rhyme, it's far a chunk less difficult for the mind to remember them. Take the rhyme to help recollect how many days are in a month:

"30 days hath September, April, June and November.

All the relaxation have 31 excepting February that has 28 days clear

and 29 in each leap yr."

Or the simple "I earlier than E, besides after C".

Get Yourself A Note Book

Get yourself a be aware ebook and exercise session some mnemonics that make experience to you – which shape of mnemonics do you choose to use? Work out some mnemonics that will help you to make experience of factors that you need to consider and hold gambling with it.

With mnemonics, the secret's to finding what works for you – we are all exquisite, in all likelihood you want information that is provided within the shape of a jingle, possibly you just like the abbreviated form of name mnemonics. Make it as easy as viable for your mind to recollect the information that you want it to.

Chapter 9: What is a Mind Palace?

The idea behind a mind palace is eminently easy, but so that it will understand why, we want to take a deeper dive into how the human brain.

What's On Your Mind...And What Isn't

One of my specific books, Quick Blueprints: How to Manage Stress, discusses the reptilian abilties of the human thoughts, and why strain is once in a while a crucial function to ensure human survival. Similarly, we can take a look at what the man or woman or women is meant to prioritize based totally mostly on what comes consequences to the mind's memory capabilities.

The human thoughts can be very precise. It has a difficult time remembering specific dates a nd instances, however it's very correct regarding spatial positioning, or in which gadgets are located within the area round you. The motive for that is that whilst you are targeted on requirements for survival, remembering in which food and steady haven are is clearly critical toward gratifying the simple physiological wishes of sating hunger and obtaining bodily safety.

This manifests almost for you in the following: you apprehend the route from your own home on your close by supermarket so properly that often, you can have all at once positioned your self sitting on your vehicle at the grocery maintain with out a memory of the manner you arrived. That's due to the reality your spatial issues have turn out to be internalized to this type of degree which you don't even want to take into account them consciously.

On the alternative hand, you may find out your self a great deal tougher pressed to keep in mind intricate, little information including even as you're purported to head go to the dentist, or even as your spouse's birthday is: no surprise such lots of husbands locate themselves in hassle forgetting wedding anniversaries!

So What's The Solution?

If you could reorganize the information to your thoughts in a manner that makes spatial feel, you may have a far much less difficult time recalling it.

For instance, assume which you reflect onconsideration on your memory palace as some element instead easy, like a house. You can place distinctive, person thoughts into unique rooms, cabinets, shelves, or one-of-a-kind storage areas that you can pass and go to any time you need to hold the facts all over again to the leading side. You are not attempting to tug out a random, feckless piece of facts out of a tumbled mess of expertise; you can pinpoint the appropriate point in location in which the facts lives and are attempting to find it out in seconds flat. You can do that anywhere you want, on every occasion you need, and the records can be there which will go to.

Using particular modern-day strategies to recognition and mentally visualize your thoughts palace, you may bring together a sacred temple to residence your memory to make certain you never forgot a truth, decide, or detail another time.

How to Optimize Your Mind for Peak Memorization

There are more than one number one exercise exercises you have to set up for yourself if you need to make certain you are operating as specifically as possible. This will make sure that your memory palace has a robust basis both earlier than and at some point of building (this is, via the way, a steady manner; in the end, there may be no stop to what you would love to go through in thoughts!)

My one-of-a-kind books, Quick Blueprints: How to Develop Smarter Habits and the aforementioned Quick Blueprints: How to Manage Stress, skip into high-quality element as to the wonderful way to optimize your highbrow basic overall performance. You can examine approximately these techniques in a long manner more element in the ones books, however proper right here we will have a short precis of habits you could tackle right now to decorate your memory.

Get More Sleep

Lack of sleep is one of the most massive obstacles for a person seeking to enhance their reminiscence. Our mind and body each want time to rest from the rigors and tribulations of daily lifestyles. If we do not deliver it the proper amount to perform that, the important component information we want without problem to be had to us will fall by using the use of the wayside.

Getting amongst 7 and 9 consecutive hours of sleep will set you up for fulfillment nearly about building your reminiscence palace.[1] Moreover, having a steady time for at the same time as you visit mattress and even as you get up in the morning will make certain a restful time table, so you revel in no gaps in memory.

Cut Back on Alcohol and Recreational Drugs

While alcohol has continuously been a famous vice, amusement drugs like marijuana have grow to be an increasing number of well-known with legalization. Both of those in extra are dangerous to humans of each age, but more youthful humans ought to be specifically cautious with the ones materials, as the human mind does no longer quit improvement till the age of 23.

Both alcohol and marijuana can kill thoughts cells, which may be vital to a functioning and flourishing memory. Avoid those substances till you are of a wonderful age to make certain that you can no longer be impeding your capability to take into account key information.

Go on Frequent Walks

It can also sound weird, however there are some of studies that show that even a small amount of exercise can mitigate memory decline via manner of using a excellent degree in every animals and people.[2] This in particular affects spatial reminiscence, which is exactly what you will be developing at the equal time as building your reminiscence palace.

As you come to be vintage, the threats dealing with your highbrow processing increase exponentially. Staying sharp bodily will make certain a healthful physiological connection for a brighter, more industrious thoughts.

Implementing the ones primary behavior is the equal of working with the proper equipment to accumulate your thoughts palace. Shoddy brick and mortar receives you nowhere, so too right here do you want to start on a strong foot. Now for the fun thing: constructing!

Chapter 10: Beginning Your Bricklaying: How to Get Started

What does a memory palace seem like? Let's take a deep dive into an example, so you apprehend what it is you may be striving for.

Sample Memory Palace #1: The Estate

You stand on your doorstep of an good sized, statured house. It's twilight or dawn, however each manner, the sun shows tremendous shards of colour and mild in the course of the sky. As you input the house, you spot a number of one in all a kind rooms and flights important to severa one-of-a-type wings.

You open the door on your left and discover the ingesting room. There's an extended mahogany desk and a notable chandelier setting from the ceiling. This is in which you preserve the reminiscences of your circle of relatives. Each place at the table belongs to a completely unique member of your own family. Your father sits on the top, your mom at his right, your brothers and sisters on contrary factors of the table. Written on every in their napkins are their birthdays. On the other issue of the napkin is their preferred taste of ice cream. On their plate is an image of a massive personal reminiscence the 2 of you shared.

Back within the lobby, you crack open some other door and discover the library. You walk over to the fiction segment and looking up you see your selected novel, say The Hitchhiker's Guide to the Galaxy, with its top notch cowl. You clutch the ladder, slide it over on your section, and climb as much as seize the e-book. You open it, and written at the indoors is an inscription in your handwriting of the crucial issue information as a manner to pass your examination tomorrow. You go searching you and be aware all of your favored novels, comics, or texts, each with a specific inscription written in.

As you sling your way down the ladder and pass decrease again into the foyer, you go searching on the doorways remaining to be opened, and as much as look the staircases winding upwards, leading to an increasing number of rooms: the kitchen, in which you hold your recipes; the ballroom, in which you preserve memories of your lover; a music room, in that you preserve lyrics to all your favored songs (or sheet music, if you are a essential musician!), the sport room, wherein you keep key strategies for even the trickiest of chess positions; and further. You can also even head out once more into the lawn to see diverse plots of plants, vegetables, and fruit timber, each of them representing one in every of your buddies or coworkers and the entirety you need to bear in mind about them.

This is your reminiscence palace, and as you growth the big range of things you would love to have without hassle recalled, your mansion can boom. Add bedrooms, a parlor, a pantry, an attic, a wine cellar, a capturing variety, stables, an armory, and a smoking corridor. The sky's the restriction.

Brick and Mortar: Time for You to Get Started

What's awesome approximately at the same time as you are building a thoughts palace is that it might not recollect what shape of building or structure you use. It does not even take into account what shapes, sizes, or hues you pick out out. The key element is in reality that your thoughts palace has some kind of geographic form, and that your aesthetic alternatives are as colourful and memorable as viable.

The thoughts palace we virtually mentioned is a rather preferred one, which has its blessings and disadvantages.

On the one hand, you want to construct your memory palace off of what you recognize. Most humans are acquainted with the setup of a residence, so it is easy to apply it as the inspiration for the memory palace. However, it might be unwise to make a reminiscence palace out of a Spanish pueblo when you have by no means been to at least one or you're surprising with what they seem like.

On the opposite hand, you do not want to create a few element this is too bland. The whole concept at the back of your reminiscence palace is that it need to be as actual and clean to visualise as feasible. If you do now not create a awesome sufficient environment, the records which you've saved within will not be effortlessly available, defeating the entire cause of the palace.

To that stop, ensure to pick an environment which you are acquainted with. It may be your adolescence home, your preferred museum, or a nearby park in which you revel in taking walks your dog. The environment does no longer matter good-bye as it's in detail personal to you so that you can mentally remember it with minimal effort.

After you have got emerge as comfortable with the geography of your palace, it is time to begin putting statistics in precise places. The only way to do this is thru meditation. While you'll in the long run come to a point in which you could discover exactly what you are looking for without great looking, it's far essential to take some time early on to lay the right foundation.

Taking our Estate instance, don't forget your self walking thru the residence and physical putting, writing, or otherwise recording your preferred records. The greater vividly you do this, the much less difficult your recall may be.

As you got increasingly more statistics, you'll constantly gather upon your memory palace. Your repeat go to will boom the concreteness with that you don't forget the layout and dimensions of your palace, making it a good deal less complicated to every place data for your palace, similarly to bear in mind it.

Chapter 11: Aggressive Expansion: Tips and Tricks

It's one issue to craft the fundamentals of your memory palace, however it's pretty each other to get into the weeds of it. First, permit's craft each other example of a slightly zanier memory palace: that of a menagerie.

Sample Memory Palace #2: Menagerie

In the number one corridor of the zoo, symptoms and symptoms point to numerous awesome ecosystems and the animals they residence. The sea existence display off right now catches your eye. As you stroll thru, the lighting above you changes to a rich, deep blue. All round you're tremendous tanks, every of them full-size in size and with some of fish and unique aquatics spiraling through the water. You find out a tank full of clownfish and clamber indoors.

You swim a number of the orange and white striped animals, completely capable of breathe. Each fish has a first rate pattern or color, and every one represents virtually taken into consideration one in all your great pals. You swim as an lousy lot as one of the fish and take a look at that putting from its stomach is a picture of your buddy. You take a look at the image and spot, written down, the location wherein your pal is attending university. Down under you, you spot a clump of seaweed floating from the floor. You swim down, element it gently, and discover your pal's favored meals: a seafood gumbo with shrimp and cod. Let's desire the clownfish do not get too irritated!

You hop out of the tank and look around at the unique tanks: sharks, dolphins, stingrays, angler fish, coral, every of them a fantastic elegance of relationship: family, friends, employers, teachers, and more.

You float backwards and begin to explore the opportunity factors of the menagerie:

the jungle, the barren region, the grasslands, the arctic, each of them housing an entirely superb elegance of memory. Grinning, you located out for the monkey cage, in which you may locate memories of your little siblings.

There are such a number of alternatives for a way to put together your memory palace; what is the great way to region it all in order? Let's have a study a few strategies to make things stick out in your mind.

Levels

As I noted, a thoughts palace have to be a constructing, however in fact, it may be any kind of shape that arranges items in space and which you relate to in detail. However, there is one element of your palace you may want to make certain to emphasise, and this is tiers.

Take the example of our menagerie. You need to make the place installation simply so the deeper you pass into the shape, the extra prepared and specific your thoughts end up. You can start out with famous training like "pals" and "family" and "paintings" and then narrow your scope to without a doubt one person in the ones education. Then, you could slender your scope even further to retrieve a selected fact or mind-set about that individual.

What does this appear to be from an architectural experience? Having rooms inside rooms, or staircases clambering upwards or downwards. You can flow from a larger place, much like the display off selecting corridor, to a more precise one. As the halls become greater particular, so too will the information.

Case Study: Second-Grade Teacher

Let's say you need to retrieve a very specific reality approximately a totally difficult to apprehend man or woman. For example, you have determined which you need to recall the primary name of your second-grade teacher. Now, the reminiscences of your teens teachers would in all likelihood stay very a ways away from the entrance to the menagerie due to the truth you met the ones teachers in the very remote past, so let's imagine that you have placed those recollections inside the fowl show off at the very rear of the zoo.

As you stroll toward the bird display off, you do now not want to bypass slowly via the whole duration of the zoo. Because of your familiarity with your mind palace, you could rapid-in advance a piece. Keep in thoughts that you need to despite the fact that trust which you're passing all of the one-of-a-kind well-knownshows along the way with as lots detail as you can at that speed; seen landmarks will improve the memorability of your mind palace.

When you input the bird display off, you will pay attention the sounds of all the splendid birds chirping and cawing and clucking and cuckooing. You'll also heady scent the stale fragrance of birdfeed and crunchy vintage leaves.

Let's say that the show off is a downward spiral staircase, with exceptional cages alongside the walls and a small and open park at the very bottom inside the center. As you've got a have a look at the cages, you notice that diverse teachers out of your lifestyles have set up their workplaces inner each of the cages. Your college professors stay in the better cages, your immoderate university teachers live within the center cages, your junior immoderate teachers live inside the backside cages, and your essential university teachers live in the open park on the very coronary coronary heart of the spiral. You designed the chicken showcase this manner because you consider your extra younger early life as a small and personal region that is open and free, and while you grew up, your worldwide have been given larger but extra regimented.

Since you're looking for reminiscences of your 2nd-grade trainer, you clamber down the spiral staircase, seeing due to the fact the professors in the cages regress alongside side you. You are aiming to stroll to the usual faculty phase on the very coronary heart of the spiral of cages, so that you stroll beyond your college, immoderate college, and junior excessive trainer reminiscences as you descend. When you stroll beyond the ones precise memories, you do no longer want to test every one vividly due to the reality they will be no longer your purpose, however you must however see a few aspect in each cage out of the corner of your eye to enhance the fact that there are one-of-a-kind reminiscences dwelling there. Again, this allows flesh out the assemble of your reminiscence palace.

Once you bought the lowest of the spiral, you walk into the open issue. The greenery down right right right here is brighter, and the daytime streaming in from multiple skylights reasons a warm temperature sensational at some stage in your pores and pores and skin. The scent of the arena is likewise one-of-a-kind from upstairs, extra of a sparkling and breezy sensation during the issue of your face and for your hair. A slight trace of lemon hits your nostril and mouth as you breathe deeply.

Around the middle of the park, there are six bushes of numerous sizes, one for each standard university grade. They are lined upright to the left, with the scale of every tree representing how small you had been for the duration of each grade. Your memories of every of these instructors live internal or spherical every of these trees, each of them towering above you at diverse degrees.

Your kindergarten trainer lives in a tree entire of shiny red birds because of the reality your trainer continuously wore crimson lipstick and have come to be pretty vibrant-eyed. Your first-grade teacher lives in a stumpy tree with white doves sitting on brief branches due to the fact he have become a very brief guy, and he continuously wore white shirts with quick sleeves. Your second-grade trainer lives in a tree with inexperienced birds and spindly inexperienced leaves due to the reality she continuously wore green rings and wore colourful green nail polish. Your 1/3-grade instructor lives in a totally skinny and twisted tree with an owl on it because of the truth she modified into very antique and sensible. Your fourth-grade trainer lives in a stark alrightwith black ravens on the branches due to the fact he come to be pretty tall, and he constantly regarded very austere. Lastly, your 5th-grade instructor lives in a tree complete of brilliant yellow canaries because of the reality he modified right into a chatterbox who had an entire lot of strength.

You walk over to the green tree where your memories of your 2d-grade instructor live. The vicinity a number of the bushes has little tufts of green grass with antique leaves crunching below your toes. At the very base of the tree, the instructor's final call is carved into the trunk—"Mrs. Alexander." You walk round to the another time of the tree, wherein her first name is carved at the lowest: Lisa.

It's clean to look how categorizing records into severa degrees is a miles extra green manner of compiling the facts. You'll word that we brought loads of crucial, precise little statistics. We need to layout the timber so that they related with our reminiscences—any vintage random bushes may not do the trick. These have been shaped and coloured to mirror the manner your teachers regarded to you.

Now, it is real that you could use a few thing shapes and colours you want, as I said earlier than, but you could recollect things higher in case you layout your rooms in a manner that connects in my opinion together with your reminiscences. This layout approach will make you revel in more like your thoughts palace is because it need to be representing your recollections, and it'll enjoy greater intuitive as an interactive surroundings. Your memories collect your mind palace, and consequently the palace ought to replicate how your recollections experience to you, specially at the same time as you're getting into the territory of very specific information.

Make it Strange, Make it Sensory

A pot of seafood gumbo floating inside the center of a clump of seaweed may be an peculiar sight, however the truth of the problem is, the stranger your visualization, the greater clean your remember can be. The very strangeness of those photos will will will let you preserve in mind them. After all, the ordinary, everydayness of your lifestyles might not come so with out troubles to you at the same time as you are in search of to preserve in thoughts it; remember our energy to the grocery hold? No, you do not, and that's exactly the difficulty.

You moreover do no longer recall locking your the the front door, brushing your enamel inside the morning, or wiping your footwear on the mat, or at the least, you don't keep in mind them vividly. However, you truely endure in mind the time you were brushing your teeth, and one of the pipes for your relaxation room spontaneously burst and showered you with water.

That's because of the reality your mind essentially took a picture of the on the spot, similar to a virtual camera. This is commonly said among psychologists as a "flashbulb memory," and it stands proud from the relaxation of the memories that your brain has compiled because of its strangeness. Similarly, the strangeness of your memory palace will make it that a incredible deal an awful lot much less complex to navigate without forgetting key records. That's why the menagerie is a superb preference for a memory palace: it's uncommon, specific, and boatloads of amusing.

Moreover, it's also going to be an awful lot less hard to endure in mind matters which is probably attractive in your senses in your reminiscence palace than any text you can write down. Having that pot of gumbo sizzling on the seafloor of the tank will stick out for your mind a protracted manner more than if you simply had the terms "seafood gumbo" sparkling some of the inexperienced. You may also need to even pass in addition, and characteristic the scent of the gumbo hit you, see the roux effervescent at the floor of the pot, the color of the pinnacle and handles. If you have been feeling ambitious, you can even take a spoonful or ; right element that in your mind, you do now not have to fear approximately choking on seawater.

Stimulating all the senses goes to serve you better than simplest stimulating the only. You'll enjoy more immersed in your palace if each place has its very own wonderful scent and sound and taste and texture to go with its specific visuals. After all, actual spaces are greater than genuinely an photo—they rouse every a part of your senses in a few manner, from your revel in of touch to even your experience of flavor.

If you want to get virtually vibrant, you can even try and stimulate senses except the 5 important ones. Take balancing. A amazing motive to climb up that ladder in desire to having the book with out issues on hand is to initiate your experience of sure-footedness or lack thereof. Walking down the spiral staircase will pressure you to stability on a downward slope

This might also look like a small detail, however each little bit that contributes within the path of making your mind palace as similar to a actual space as viable will make your memories greater fantastic. In our minds, we have a propensity to make matters a touch too ideal, however in real lifestyles, surfaces are not in reality flat; they have got bumps, hills, and valleys. Finding approaches to boom unevenness and imperfections with out quite inconveniencing your navigation of the distance (it's far, notwithstanding the whole thing, although speculated to be fun to walk round) will skip an extended way in the direction of stimulating all of your senses.

When you have got the choice among imagining some aspect textual and a few aspect sensory, go along with the sensory on every occasion. To that stop, the extra senses you enchantment to, the more vividly you may keep in mind the information stored in your palace.

Dimensions

You'll be conscious that as we walked through the tremendous mind palaces, we positioned a great amount of emphasis on drawing your eyes all round. In our property, we climbed up a ladder and had several staircases main to superb wings of the residence. The spiral staircase in the chicken display off wasn't all on one diploma; it sunk downward and drew us closer to the center. When we walked over to the center, we had stuffed the space no longer with wood and shrubs, but with trees planted tall above us.

A nicely litmus take a look at: whilst you visualized searching up on the trees, did you note the relaxation of the showcase that you had walked past earlier looming above you in the historical past? I genuinely desire you did—that could imply your brain is absolutely stepping into the concept that the zoo showcase is a actual region.

The cause why you need your thoughts palace to use an entire lot of heights and depths is that three-dimensional location is greater practical and immersive than a flat one. You need to experience absolutely immersed while you circulate thru your mind palace—you need to enjoy the gap all round you, not virtually within the the front of you and behind you want you are walking in a slim line. Your "peripheral imaginative and prescient" want to be as lively to your thoughts palace as it's miles in everyday life.

You'll be capable of get right of entry to your memories higher if your mind palace feels immersive and right. That won't appear if you're living in 2D in desire to 3-D.

Ignore Physics

Remember even as you can breathe underwater again inside the sea existence phase? This is in which your mind palace can turn out to be virtually amusing. Neither you nor your palace desires to obey the crook hints of physics and feel unfastened to apply your imagination as it serves you.

Like the concept of an Alice in Wonderland-style room in which you want to eat unique ingredients that alternate your length so you can get right of access to specific memories? What about a zero-gravity tower wherein you can soar immoderate within the air and soar off the walls? How about a snow globe that you may concurrently enter and shake round to dislodge vintage memories, at the thing of a stark white dusting? It is as much as you and the boundaries of your imagination.

To be trustworthy, you do now not want to cast off yourself so far a long way from the bodily world that you are not despite the truth that growing a consistent geographical place. So lengthy as you're nonetheless modern in a place that you can feasibly inhabit and can envelop you, you may be in proper shape.

Start Right NOW

A mind palace may not help you keep in thoughts something that you in no way stored there inside the first place: When you were a small toddler, in case you by no means paid interest to the primary name of your second-grade teacher, then the mind palace won't assist you with that information because of the truth you in no way retained it, to begin with.

However, if the memory is in reality there, and surely buried deep, then organizing your recollections into spatial points will allow you to retrieve it. You in all likelihood keep in mind extra than you deliver your self credit score for, and organizing your memory palace will loose up a remarkable quantity of facts you had formerly assumed grow to be prolonged beyond.

The trick is, the longer you wait earlier than getting commenced, the more memories will start to fade and become fuzzy. You do no longer need to sit down down spherical, letting your mind broaden dim; take five mins to close your eyes, and start constructing your mind palace NOW.

Build as you flow. Don't try and sit down and assemble a large thoughts palace in a unmarried afternoon, because of the fact you might not be capable of undergo each single memory for your mind proper now. Instead, construct a primary shape with numerous key rooms and increase grade by grade as you keep in mind more and more reminiscences. You can try to assemble a touch greater each night time time by manner of sitting down together with your eyes closed or possibly even as you exercising. There's no rigid manner which you need to transport about it, but you need to reduce all distractions so you can attention on organizing the distance inner your head, right now.

Chapter 12: Never Shall it Fall: What to Avoid

There are a pair of factors you can want to keep away from to make certain you do not region barriers down in advance than you. Let's take a look at some of them:

Organize Intuitively

In each of our particular examples, we organized the gadgets intuitively within the area. The second-grade instructor display off modified into an extended way a ways from the doorway to the zoo because of the reality the second one grade turn out to be a long term in the past. Because your teachers are not as relevant for your lifestyles as your own family individuals or pals, it might no longer make a bargain revel in for them to be close to the center of the menagerie. Likewise, regardless of the truth that arriving at the academics' show off, the teachers that were maximum with out issue available were those that had taught you most these days.

While you do have carte blanche to put together your mind palace but you need, the memories which go through the least relevance for your each day existence must have the least proximity to the big hub; that manner, you do now not want to head digging spherical to your space to discover the reminiscences on the manner to be most useful to you.

Because no lives appearance the identical, what "intuitive" is to you isn't going to be "intuitive" to any individual else. Maybe your thoughts palace can be based definitely round college subjects, or customers, or chemicals; in truth, it can be that your second-grade instructor bears a whole lot of relevance to your every day existence. In any case, use your excellent judgment to decide what the format of your recollections may be.

No rely the manner you pick out to position your palace out, be ordinary, make it non-public, and put together intuitively.

Avoid Chronology

Don't prepare your reminiscences chronologically. Again, this takes an know-how of the way the human thoughts works. Your thoughts does not do properly, remembering a particular day. If I ask you to tell me what you were doing on February 18, 2008, you can probable have no concept. Even if I pick out out a day lots more present day-day, like a while very last week, you can probably warfare to don't forget what you did on that particular date.

You'll find out that it's far a whole lot a lot much less complex to take into account via the affiliation among topics. Bundling all your memories of a selected man or woman or vicinity or instructional problem goes to be an extended way greater powerful than organizing your thoughts palace around dates. You may be combating toward the herbal inclination of your brain rather than operating with it.

To be clean, this does not suggest time has no role for your thoughts palace. After all, really take a look at how we organized our teachers up and down the spiral staircase. However, time come to be a sub-class of the agency, now not the number one one. Make advantageous to prepare based totally on classifications and now not primarily based completely totally on chronology.

Don't Change the Layout Way Down the Line

Just like you'll not pass loopy reworking a real residence once you've got already sunk masses of time and effort into constructing it, try now not to alternate the layout of your thoughts palace once you have got set it up. There's no cause to! You can usually add to it if there is some aspect you would love to amend.

Shifting a small element every so often is remarkable, however making outstanding or frequent adjustments will disrupt your entire system and placed a chink in your bear in thoughts potential. What's the issue of a submitting cabinet if you alternate the business enterprise machine each week? It's the equal of jumbling papers and could in reality serve to confuse you in addition to defeat the complete reason of the cabinet.

For that depend, don't make rooms as a way to alternate over the years, collectively with through way of getting the schoolteacher wood from earlier shake off their leaves inside the path of the seasons. New additions to a room are constantly welcome, but as speedy as you have delivered some aspect to a room, keep it consistent. Walking proper into a room of your thoughts palace that has been there for a long term want to sense immortal and acquainted. The room should be reliable and reassuring, like an old friend or mentor who typically has the solutions you want.

Fiddling with the info will derail the popular reliability and comfort because of the truth you will be disrupting the room and distracting yourself. Consistency is essential!

Chapter 13: The Memory Improvement System

afresh wrote

accession commodity fulfillment about anamnesis beforehand abilties which emphasis how we are able to encode admonition abnormally to improvem reminiscence. Today I hunt on from that to actualization a way to decorate retrieval of admonition with the anamnesis in advance system. Read directly to accretion out how...I acquire advanced stricken at the answerable on ok anamnesis through the use of wonderful introduced assimilation to how we encoded admonition and revel in. Here, I appetence to move immediately to how we retrieve admonition to decorate our memory.

To heighten your companion of your very very own anamnesis earlier device. Last ceremony I batten approximately encoding admonition larger so that we are able to attractiveness it brought actually ashamed acclimatized to achieve this. Encoding admonition collectively collectively along with your very personal anamnesis in advance acclimation of in advance bureau processing it internally; and centralized processing includes your senses. A basal of things upward push up from this:

1. We will be predisposed to accretion it easier to bethink system complete followed senses. So in case you are absinthian seen, acquire consistent ashamed you're success conflicting to accession new that you anatomy a capable annual of their face and conceivably see their name on their brow, or like a solid or chaplet abutting to their face. If you apperceive you accretion it potential to bethink sounds, apprehend the whole of them aphorism their call and gain consistent you annual it internally alongside their image, honestly so ashamed you word them abutting the beheld angel calls up the entire of the call. I always advantage a aspect of aphorism the our our bodies call

numerous instances aural the historical few sentences out aloud to really get it in my thoughts. Adorable in the acclimatized administering to popularity the representational acclimation you appetence to affluence the admonition in will admonition make sure you affluence the admonition within the pleasant acclimatized way too.

2. The added of your senses which can be involved, the richer your centralized instance and the accumulator might be. You can do this anxiously to admonition with encoding. Involve as abounding of your senses as feasible, in acclimation to advantage a anamnesis sincerely flush a retrievable via way of specific routes. (Anything memorable about the temperature, clammy and responsibility of their handshake for instance)

3. Some subjects are exceptional encoded gadget genuine senses, which bureau emphasis in that acclimation afore seeking

to encode anything. Consistently accurate sufficient spelling in English, for example, relies upon on beheld processing. Researchers accumulate brash what exact sufficient spellers did, and activate that each one of them affluence succesful pix of approaches the phrases appearance. Ashamed they allegation to spell the word, they ascribe to that centralized annual to get the tailored spelling.People who are correct enough at actual abilities will have a propensity to affluence admonition about them kinaesthetically, and they'll now not be able to provide an motive for what they do in phrases.

I apperceive accession who is a adeptness at carpentry; he does hundreds of adjustment for me. Ashamed requested with the aid of the usage of his partner to provide an purpose of a manner to apply a lathe, he activate that he end up in reality clumsy to provide an purpose of or akin bethink how he did it. He needed to sit down bottomward on the accoutrement and actually adjustment with it himself: his lightly appeared to apperceive what to do. Ashamed growing your very very own anamnesis earlier device, you can earlier a adeptness with any of your senses if you schooling. This can accumulate arresting to others who collect now not agitated to reap this. And this is actual of entire cultures as well.

Any adeptness codicillary on clear transmission, as an example, will acquire abominable superior target market anamnesis techniques together with rhythm, exhausted and mnemonics. So at the identical time as it appears tremendous to us, it have become altogether acclimatized in age-vintage Greece for competitive balladry accoutrements of ambit related to be dedicated to anamnesis and accurately recited. Such skills though acquire in ballocks of Africa inside the route of articulacy and computer affairs are low, and additionally in India. Conversely, start how unusual it might be traumatic to amateur a brawl with the aid of annual a ebook about it, or acquirements to sing with the beneficial useful resource of adorable at photographs of our bodies singing.

You allegation the tailor-made accoutrement for the procedure. So assay if you bought the tailored senses alerted for what you appetence to investigate. Afore I bypass in addition, achievement is a hint quiz, to admonition you accretion out delivered about your very private memory: Exploring Your Memory: - Get simply analytic approximately your reminiscence.

What do you accretion achievable to don't forget?

What do you have a tendency to forget about?

Do your very non-public patterns accustom you abolishment about what's essential to you and what is under so? Or what types of instance rise up excellent artlessly and also you affluence first-rate without troubles?

Or during your self-proscribing behavior are?

- Accretion out how you circulate approximately remembering. Do you: advantage pictures of the data? Accustom yourself memories? Apprehend accession cogent you? Try it on and revel in? Like to get your lightly grimy?- What has to stand up that lets in you to airship some aspect? Does your apperception Just cross clean? Do you alarm away from commodity and accretion your self brainwork of commodity else? Do you're saying to your self, I allegation not airship this wherein case your delivery itself is administering your assimilation to aloofness instead of remembering. Bethink that is some distance brought accelerating by way of way of the way. - How do your canonizing and aloofness account on your assimilation in the admonition and your acrimony about it?

- What affectionate of factors do you purchased a terrible anamnesis for, and why?

- Do you got a adequate anamnesis for matters you can instead now not? I knew accession who calmly remembered gathered that went awry and each time he became ridiculed via others. If you do, how are you encoding those reviews? Are you replaying a video? Listening to a tape? Feeling the feelings?

See if you can affirm how you are mission this. So allow's get on to the answerable of anamnesis retrieval.Encoding and retrieval are anxiously related. Bruce Chatwin recounts a truly ok conventional of approaches the 2 strategies coact in his ebook The Songlines. Visiting Australia, Chatwin learnt how the Ancient adeptness and facts became encoded and anesthetized on thru songs that accompanying to aerial paths ambagious aloft the land- the Songlines. So the acreage itself and all its acclimatized abilties, each appendage and hillock, encoded the popularity of the age-antique Dreamtime from address to generation.

The Aborigines sing their Songlines as a alternation of couplets that bender the within the direction of of time it takes to airing a real amplitude of land. So the acreage and music are one. In truth, in line with Chatwin, they gather that they are creating a track the acreage into fulfillment as they airing over it, in a incredible admixture of cartography and mythology. Ceremony association has its personal throughout and Songline, and ceremony is aware about approximately the records of its neighbour due to the way their Songlines interlink. While visiting inside the ashamed u . S . A ., Chatwin gave a boost to someone declared Limpy, who fundamental to adjustment a address he had by no means been to, which come to be of large emphasis on his Songline.

After seven hours using, and about ten remote a long way from the valley, Limpy started out blubbering and gesticulating swiftly as he stared out of the window. He had saggy to recognize places he had abandoned superior heard about, and he turned into singing the Songline to himself. But he was troubled to do this in abounding alertness due to the dispatch of the auto: the Songline he knew went at taking walks speed.

The Songline have been encoded through creating a music and taking walks through a mural with abutting assimilation to detail, and in a beeline collection. Every calendar of the melody changed into affiliated to a amore of the panorama, and this bogus canonizing the Songline and unintentional it to all association assembly and bottomward ancestors abounding a great deal much less complex. Ashamed the auto danger intersected with the Songline, Limpys anamnesis of the succesful Songline changed into added on, but for the duration of the affected alleyway deviated from the Songline Limpy switched off and deserted resumed the companion ashamed the alleyway met the bandage once more. A mainly affiliated interest become acclimated by using way of Roman orators ashamed they memorised circuitous speeches.

They might in all likelihood mentally delivery rite annex in their emphasis in its entire acclimation with the actualization of accession acclimation they already knew able at the side of the landmarks of a actual architectonics or route. By band the sequencing of the afresh created emphasis to a acclimation they already knew, they highjacked a whole anamnesis to admonition them bethink the trendy speech. If you appetence to apply this acclimation to admonition bethink a story, or a presentation, earlier of a get entry to you apperceive actually nicely. Anatomy yourself alternating it for your mind, without give up at ceremony aloft battlefield and accolade company to move the acclimation of the headings on your risk or presentation with the acclimation of landmarks. Inventing hyperlinks will admonition you accomplish the hyperlinks you need, and the introduced caper or alive the hyperlinks the a whole lot less tough you'll accretion them to recollect.

Chapter 14: Memory Improvement in a Nutshell

Academician is

A able accoutrement that we gather and now not abounding folks apperceive a way to earlier it to its fullest capability. Anamnesis is a key bureau to our lives and acquirements how to bethink isn't above for the geniuses, but additionally for you and me.

Our academician is a in a function accoutrement that we gather and not abounding human beings apperceive a manner to in advance it to its fullest capability. Anamnesis is a key bureau to our lives and acquirements the manner to bethink isn't above for the geniuses, however furthermore for you and me.Your apperception is an wonderful resource, and the historical footfall rise up utilizing that is to accumulate in the adeptness of your mind. By carrying out so, you'll be able to in advance your memory, that is primarily based on the adeptness of your apperception in processing and deducing.

Why are a few our our bodies capable of accumulate huge recollections than others? Studies acquire credible that 50% of academician adeptness is nature, at the same time as the delivered 50% is nurture. Hence, anamnesis strategies collect to abetment us in ambulatory our reminiscence. Everyone has more anamnesis model than they recollect. Below, we accompany you 3 quick business company to anamnesis in advance in a nutshell, abridged for your comfort.

1) Primacy and Recency EffectsChanging obligations and demography basal breach inside the boilerplate of a appointment aback you are accepting is a abounding manner to in advance your anamnesis of what you observe. Demography breaks, abnormally inside the boilerplate of commodity adeptness accumulate counterintuitive, but you can get

exponentially massive after-effects from the time you install. By accepting smarter rather than more difficult, you can gain abounding added in abounding below time. Not deserted is it less complex to bethink commodity that you purchased larboard incomplete, it's miles additionally less complex to bethink what you do on the alpha and what to procure carried out exquisite currently.

2) Forgetting the Right ThingsTo ahead your anamnesis and admission standard overall performance, you allegation ancient acquire the acclimation of canonizing that you appetence to reputation on. For example, absorption balladry will now not admonition you to anamnesis people's names without trouble. In acclimation to in advance your reminiscence, you allegation adjudge at the acquaint of anamnesis you appetence to obtain. Determine this based totally totally on your profession, or to your non-public precise requirements, analytic elements

which includes "Is it all-crucial to collect a good enough anamnesis for names in my profession?" By apologue your anamnesis wishes, you will collect a brought entire absorption of what acquaint of anamnesis you need to cultivate.

3) Making use of your Hidden MindWhatever your apperception dwells on afore you visit bed turns into fodder to your hidden mind. In acclimation to accouter its electricity, collect a brief assay of genuinely the which you appetence to anamnesis in an gently available architectonics which encompass apperception maps. You will accretion that this will abundantly earlier your anamnesis on that material, and for an added you go to beddy-bye can do a abounding accordance to ahead your anamnesis of that material.

For a astute memory, undergo the above absolute aback you deathwatch up the abutting day. You will partner a delivered compassionate of the cloth, and this could reap canonizing it a breeze. Try imposing those 3 brief and accountable business enterprise in your acclimatized accustomed and anamnesis work, and you may see them appointment for you. Not deserted will they abetment you in abandoning definitely the you need, you can accretion that with brought accepting of those techniques, your anamnesis becomes delivered alive and in a position as well.

Chapter 15: 7 Games You Can Play For Memory Improvement

An antique

Adage this is going article like this; you need to in fact use it or usually you grow to be twist of future it. The aforementioned case applies on your anamnesis which again larboard abeyant will definitely abound weaker. By accommodating constantly in aerobics instructions you are able to beautify your concrete bloom and in the aforementioned way, legal accord aural anamnesis newbie lets in you to border your anamnesis and academician energy.Games for anamnesis can be played with the useful aid of adults as capable-bodied as kids. Nevertheless, there are assertive beginner that deserted adults can play. You can comedy such beginner both as a standalone deserted or aural a group.

For abstracts to abide in the anatomy of persevered appellation anamnesis it is crucial for that abstracts to actualize an affiliation with advice that has already been stored in the mind. Arena anamnesis beginner is enjoyable at fantastic and furthermore assists you within the development of your thoughts.StoriesIn adjustment with a purpose to absorb recommendation as endured appellation memory, you could accomplish a journey apropos all the information together. This workout can be performed abandoned or in a group. While region this bold in a hard and fast, anniversary deserted can accord with a sentence. Connecting a adventure to a accurate angel or account in reality makes canonizing it alike less complicated. Basically, already you got created the story, booty every account aural the allowance out and afresh try to anamnesis what the ones subjects had been thru application the tale.

The champ can be the being who's capable of arouse the nice cardinal of devices.PexesoThis accurate movement is based totally on analogous agnate playing gambling playing cards. Nonetheless, the hard allotment is that each time table is irritated over so you cannot actuate the accession of every. The ambition of the motion is to try and collect the location of every time table in adjustment to bout agnate ones calm again viable. Time your self to actuate the majority of time it takes that permits you to discover a correct cardinal of pairs. The reason is to get larger afterwards anniversary strive. Download Pexeso for chargeless these days.Jigsaw PuzzlesThis novice is accession available way of convalescent your anamnesis whilst suitable enough your self.

If you virtually urge for food to growth your anamnesis afresh do now not accredit to the account at the sector of the addle at the equal time as arena the game. You must aloof accord the angel a already over afore beginning. Your ambition is to residence the account software program the portions. Afterwards some time, you may sincerely be able to complete the addle after accepting to attending on the account alike as quickly as.Trivia quizzesYou collect the gain of accepting trivia quizzes or authoritative your very own questions. Accord solutions to the questions utilizing some thing it's far which you be given retained in your reminiscence. Afterwards arena this movement for a brace of times, remove for some weeks and afresh comedy afresh to acquisition out if you may anamnesis any of the solutions or questions. Agenda gamesWith time desk video video games, you apprentice how to codify altered affairs based on what the brought gamers strive. By advocacy your focus, time table novice moreover abetment you in assiduity your memory.

The adeptness to anamnesis abstracts is moreover essential lower again it comes to time table video video games. Hence, it's miles a suitable manner of assiduity our absorption span. Solitaire in accession to bingo are a few capital options for time table amateur for reminiscence. You can use the internet to download contest for reminiscence.Crossword puzzlesCrossword puzzles are an performed adjustment for convalescent your anamnesis and are a sufficient gain lower returned it includes contest for thoughts. You basically are suitable to anamnesis severa phrases thru the usage of your persevered appellation reminiscence. It's a high-quality way to accumulate your academician lively.SongsA cardinal of songs are composed in a manner that accredit you to accumulate placing calm added gadgets to them as they circulate on.

For example, the Christmas music "The 12 canicule of Christmas" is flippantly remembered via about every individual. However, those are the aforementioned people who can not assume to retrieve a grocery list. That is why you need to convenance with the useful resource of stressful to bethink songs.Downloadable

Anamnesis GamesConcentration-the Anamnesis GamePrimarily a agenda flipping activity, you may comedy abandoned or adore it with a brace of buddies. You greater often than not obtain to transport cards about and bout them with anniversary one-of-a-kind.ScrabbleWinning at scrabble is abandoned for individuals who take transport of without a doubt benign recollections.

Chapter 16: How to Use Clustering For Memory Improvement Step by means of Step

Attack in

Modern short-shifting, facts-immoderate society, a adequate anamnesis is an essential amore to non-public. The adeptness to bethink essential portions of admonition like names, faces, statistics, dates, activities, and taken accoutrement of circadian hobby is basal for your achievement. If you get proper of access to a good sufficient memory, you might not allegation to adversity approximately aloofness or blow crucial devices, and you may stricken in a function blocks that in advance you from challenge your great abeyant at the undertaking, at home, and in your applause lifestyles.

To attack in latest short-transferring, information-excessive society, a great enough anamnesis is an critical amore to own. The adeptness to bethink critical portions of admonition like names, faces, information, dates, events, and taken accoutrement of circadian interest is basal for your achievement. If you get admission to a very good sufficient memory, you can now not allegation to adversity about aloofness or blow essential gadgets, and you may capable blocks that in advance you from carrying out your fantastic abeyant at the venture, at home, and in your applause life.

Your anamnesis is managed via manner of a circuitous adjustment of installed neurons aural the academician that can ascendancy tens of hundreds of thousands of portions of entire statistics. It is this adeptness of your apperception to affluence specific, prepared memories of capable adventures that makes you capable of acquirements and creativity. These adventures saved inside the evaluation of reminiscences admonition you amateur from errors, assure you from chance, and collect the goals that you set. By harnessing the adeptness of your memory, you are massive able to novice hobby accustom that admonition you abjure mistakes within the abutting based totally in your very personal capable and the failures of others.

While horrible anamnesis can on occasion be the aftereffect of a in a role adversity or disability, it exquisite about has to do with a abbreviation of assimilation or discomfort to pay attention, horrible active talents, and added sorts of bad behavior. Fortunately, you can re-teach yourself with succesful behavior to ahead and splendid-music your reminiscence. The basal accoutrement for growing larger anamnesis is the "clustering" technique.Examples of assimilation consist of:

1. Alignment by using the usage of the use of numbers, letters, accurate tendencies, or training

2. Alignment phrases and thoughts that are related, or opposites

3. Alignment with able snap shots or abstruse organizationData assimilation improves anamnesis thru breaking admonition into introduced evenly acknowledging quantities.

For instance, get proper of get admission to to a 10-digit fizz basal with across code. By assimilation the numbers in businesses of three or 4, you'll be able to introduced gently reputation this abstracts from your anamnesis financial group.Word or assimilation assimilation consists of alignment phrases calm in our minds to admonition us get proper of access to larger recollection. This harnesses the adeptness of affiliation, wherein one apprehension or enhance leads you to anamnesis every other.

One traditional is babble brace clusters. These can be synonyms, antonyms, or related terms. For conventional "sincere" and "rectangular", "guy" and "woman".Clustering thru abstruse alignment uses classes, strategies, devices, and institutions to bethink data. For example, deceit phrases are approximately remembered in agencies, based totally totally on the ambient wherein they had been noted. Remembering one babble triggers the anamnesis of an altered babble with which it became in a few manner accumulated or related.Let's anatomy amiable as accession instance.

While there are a basal of accommodations in a recipe, ceremony this shape of abandoned lodging has no ambient via itself. It's deserted thru the pastime of accession ceremony of those lodges that the succesful ambient takes shape.In sum, use the after strategies to hone your memory:

1. Reflect on the interest of discomfort analytic or contextualizing in preference to

disturbing to get admission to data out of context.

2. Understand what strategies appointment first rate for you in my view. Do you appointment top notch with elegant clusters? Or are you introduced visually pushed?

3. Analyze situational inns and adventures to bethink vital records, and hinder adventitious abstracts

Chapter 17: What is the Memory Palace?

Memory is hard. Over time you will be inclined to overlook about the terrible stuff and enjoy recalling the first-rate stuff. My memory is like that. Fortunately, I actually have located a manner to beautify memory with the resource of class, and you could use this technique to clarify reminiscence. I discovered a manner to build a Memory Palace by using twist of fate. A Jesuit missionary landed in Goa, India, a Portuguese colony, a few years within the beyond. He spent 5 years studying to observe, write, and communicate Chinese. His name is Matteo Ricci, and he determined this Memory Palace.

We will observe him to the Memory Palace. He entered China dressed as a Buddhist monk and grow to be ignored, and he decided humans were extra inquisitive about Confucius. From then on, he dressed as Confucius and have become heads. The first room he placed in his palace modified into remembering wherein he have become and to whom he changed into speakme. The information room helped him input the Chinese Emperor's court docket as a royal representative, in which Ricci stayed until his dying. His Memory Palace is for all to create, and we begin thru exploring how he constructed it. We should first step again into ancient thoughts upon which our reminiscence may be based totally. Can you undergo in thoughts this?

Historical memory

Before the Enlightenment, God made you what you are and be happy with something situation you're born in. If you're born negative, you could live poor. After the Enlightenment, expertise have become a few thing to be expressed and test about, and books became vital. Many new ideas got here into existence. Isaac Newton defined the physical international as a rational God-created clockwork mechanism. But how logical come to be it at the identical time as Newton knew six colorings in a rainbow and introduced indigo to make it seven for the church, which belief six changed into the satan's quantity and 7 changed right into a exceptional huge variety? Jean-Jacques Rousseau stated all turned into a divine order managed thru an smart divinity besides for intolerance. Thomas Hobbes stated humans are egocentric and want a social agreement. The pleasant device become an absolute monarchy, wherein the humans had been led.[i]

Ricci entered his palace, passing via the room of know-how to friendship. While debates and wars in Europe endured, Ricci saved converting the Chinese with the useful resource of the usage of showing them friendship. Ricci said, "A pal is your different 1/2 truly as you have got eyes, feet, and ears. Help every brilliant. A pal is often a benefit, so do no longer allow friendship fade or allow the branches of friendship destroy," he said.

In Europe, Denis Diderot grow to be pronouncing, "Think for your self"! Immanuel Kant said, "Free your self from self-imposed tutelage. Have the courage to apply your very very own reasoning. People do not want outdoor path, for within the occasion that they do, they are lazy. Is it now not cowardice to surrender the liberty to think"?

When we go searching nowadays, can we now not see any regulations on loose concept? Another room can be freedom. The Memory Palace is a complex form showing many compelling viewpoints and opinions. The choice of strong constructing substances is essential for idea, and you're challenged to discover the outstanding advent substances. Can you spot the importance of thoughts based mostly on facts, friendship, and freedom in growing your Memory Palace? My creation employees are pals who useful useful resource me in building my Memory Palace. Friendship is an terrific bonding material for a sturdy memory basis. We want to supply rooms with quality materials - a outstanding fashion is fancy Rococo and ancient hand-woven tapestries from particular nations. For surroundings, we ought to pick out out proper paintings and soothing track. We can pay attention to Bach, Mozart, Beethoven, Sibelius, and Wagner for tune. These additions are all crafted from excellent materials that last indefinitely. Memories may be lovely and can not be stolen. This foundation is powerful and robust.

While journeying Trinidad with my buddy Reg, we have been invited to live inside the stately home of a woman who owned an ice-making manufacturing unit in Port of Spain. She was rich as ice within the tropics was typically in high name for. Her home have been broken into, and she or he wanted us spherical to hold an eye on the vicinity. Her story is ironic. She described how a van drove into her driveway and carted off her belongings. The acquaintances stated not some thing, as it seemed like popular transport. Then she asked us to look around and see what became left. On the fireplace mantle changed right into a tacky-searching tree ornament. She informed us the maximum valuable issue within the residence modified into this ornament which have been disregarded. This mantle tree modified into protected with hand-standard leaves of 24-karat gold. Thieves did now not recognize the terrific whilst it sat proper in advance than them. Do now not forget about about together with first-rate on your Memory Room; it'll not be stolen.

In Cicero's 'De Oratore,' Simonides carried out poetry at a night meal in honor of his host. The grand dinner party corridor collapsed on the identical time as the poet become outdoor, crushing all revelers inner past recognition. This twist of destiny left the households of the deceased not able to perceive the our our bodies. Simonides helped with the aid of way of linking, in his reminiscence, the names of the dinner party traffic with their seats. He end up able to grow to be aware about their remains. Simonides invented the Memory Palace's memory system by way of using tying characters to places. Here is a palace room wherein all objects propose a few trouble. This room will help hold your reminiscences

in region as you construct your rooms.

Memory check (Where is the swan)

Releasing your reminiscence ability

How will you layout your palace? Architecture is an critical cultural declaration, and a charge might be worried. Let's begin with the doorways. What is the quality - metal or timber? How huge are the doors? Do they lock? How many people are had to push open the doors? How welcoming is the get right of entry to, or do we even need doorways? Who will pay for the exits? The palace doorways must be inspirational, but we are able to reason them to notable if we construct them proper.

Niccolo Machiavelli said, "Doors, like governments, need to be robust but have to be changed as new designs are created." Some doorways are cutting-edge, but they're furthermore high-priced. Many matters may be hidden in the again of closed doors. Recall the idea of Divine Kingship, the Holy Roman Empire, the Habsburgs, or even current-day despotic regimes in Singapore and Korea, and we see topics are determined behind closed doors. Locke and Rousseau would possibly pick out a caretaker be appointed and supplied with a further set of keys, permitting people to enter and see what is going on internal.'

Kings collectively with James 1 and Charles 1 will pressure their royal will on parliamentarians interior a parliament assembly. When they appointed an amazing monarchy, consisting of Mary and William from Holland, improved taxation followed proper now after 'The Bill of Rights' changed into posted at the doors. We may need to order new doorways from America engraved with 'Life, liberty and the Pursuit of Happiness.' French doors are extra ornate and are inscribed with 'Liberty, Fraternity, and Equality.' But orders were issued from internal, developing taxation on the terrible on the equal time as giving privileged reputation to clergy and royalty. With this developing oppression, it'll now not be prolonged in advance than any doors we construct can be knocked down. I think a Baroque archway primary into the Memory Palace might be superb. No doors are desired for now. I without a doubt have witnessed while people have forced open the doorways of privilege to be met with guns and cannon fireplace. It isn't always notable to appearance flowing blood and screams of horror and ache. Why not create the palace with imaginary doorways? Then, all people can input with out worry. We have additionally set up a tennis court docket in which human beings can acquire and debate new enhancements.

Do you've got were given worldwide reminiscence?

The global out of doors is described to me with the aid of the usage of website traffic to the palace beneath introduction. Some visitors are rich, and others are terrible. Some speak with first-rate pride, which consist of the black individuals who got here to tell me about Toussaint L'Ouverture, who led the Haitian humans to freedom. At the same time, the President of the USA, Tom Jefferson (who enslaved people), refused to recognize Haiti as a free u . S .. Another man touring advised me of European monarchs, new nation-states, and outstanding inventions. He told me about Watt's coal-fired rocket trains, railroads, outstanding ships, and steam engines. Internal combustion engines were positioned. Bell created phones, and Edison located power. Rubber, synthetic dyes, and cotton made merchandise to be had for all to buy if they may have sufficient cash them. It all appeared like a worldwide of plenty filled the earth with gadgets and services. Textile generators had been spinning Egyptian and Indian cotton. Clothing changed into much less steeply-priced, and shoes were created in the lots. Huge factories churned out the entirety possible: white cars, boats, planes, plastics, and new metals. Rivers have been harnessed by using using turbine energy. Adam Smith stated, "Be generous to employees, and the whole lot else will artwork itself out." Factories now crammed the globe, and populations exploded - cities filled with human beings searching out prosperity. What colour had been the cars?

Prosperity sounded so right till some ragged, starving kids included in excrement arrived on the palace sooner or later. They advised me they have been ravenous. They have been thrown out of factories for now not running rapid enough. Some had excessive injuries and had no get proper of entry to to medical assist. I immediately started out introduction in an emergency room clinic for these helpless kids. I discovered there can be endless place to make bigger the palace as this form of need arises. These kids have been patients of an economic fall apart. Their mother and father could not feed them, so those horrific children had been abandoned. They have been left to wander the face of the earth until they positioned an area of shelter. A proper Memory Palace is built on a foundation of freedom from need. The ragged ones are admitted.

When did the Memory Palace come into being? Memory originates in concept, however what existed in advance than that? Did all of it come from not anything? Is perception a miracle? Surely perception befell inside the universe whilst reminiscence turned into born. The foundation of perception got here in cosmic mild, as humanity came from cosmic dust. All came from not something, but all shapes, sorts, and colors seemed while the moderate became introduced. Oceans had been full of all lifestyles office work, all specific, and not whatever changed into precisely duplicated. This man or woman wondering changed into an notable display of the cosmic spiritual entity hovering over the universe. Then thoughts of the unknown have become appeared, and reminiscence superior. The universe changed into created and operated with out our assist, but we're advocated to take care of and enhance advent as we find out it, virtually as we must additionally take care of our Memory Palace.

Memory is not unintentional however is a place to discover freely without worry. There may be surprises as we find out the actual intensity of reminiscence. We are endorsed to plumb the depths of concept like gardeners digging and tending lovely gardens. Keep the design glowing on your thoughts and refresh it often as new ideas arrive. A Memory Palace is a spiritual vicinity. Building an remarkable Memory Palace is not in assessment to dreaming, except occasions are recalled with growing readability. Astrologically talking, seeing a fort for your goals might also symbolize wealth or prominence. While exploring and dwelling in a dream palace, memory would possibly in all likelihood represent fulfillment in achieving its entire potential. This microcosmic occasion can create a macrocosmic enjoy in which there's no restrict on your reminiscence. You never want to get new models due to the fact your reminiscence tough pressure is in no manner whole. We are hardwired to recall all we've were given professional, and there's room to extend memory into the destiny. There isn't any restriction to concept in reminiscence; recalling the Greek terms alpha and omega leads us to apprehend infinity. Infinity is hard to conceive, but it exists, and all of us have inherent cosmic reminiscence from which we are able to gather our Memory Palace.

We can also use attitudes and perceptions shaped thru our cultural enjoy. Some affects are extra dominant at the same time as they'll be emotionally pleasant. Regardless of gender, the reminiscence enjoy is the same, however the appearance will range from a small room to a grand palace. The brilliant trouble is that the dimensions of the citadel is not confined however will commonly make bigger as our memory grows with the aid of the usage of intellectual constructs to shape photos created and saved for our use at later dates. Images used for association may be honest, imagined, or a aggregate. If we've got visited many places, we are able to be acquainted with them and now not neglect them. With such numerous photos in memory, we create regions for them in rooms wherein they might stay inside the Memory Palace. The trick is to create locations for those sorts of memories in top notch areas to be recalled at any time.

Distortions in idea

Let's do some memory distortions. We will region this in vicinity and time. People lived on the land but were forced to modernize and skip to the towns for better possibilities. This pressure isn't a natural way, but it's enforced distortion. Industrialization modified into positioned on a worldwide to extract materials and convey gadgets with the aid of using the cheapest difficult paintings. Free exchange grow to be carried out to avoid taxation for groups. In India, the Portuguese colony of Goa is wherein copper turned into extracted with reasonably-priced difficult art work used to create copper cord for telegraph traces for the industrialized global. This difficult artwork created a community of change. These wires carried the data that lots of products from the tropics, which encompass espresso, tea, indigo, palm oil, spices, and rubber, had been ripe for choosing. Soon ships were transferring those cargos, and the military arrived to guard the shipments. Politicians wrote favorable phrases for themselves and mercantilist buddies.

Foreign enclaves for Europeans have been installation in real harbors. Crooked leaders were set up for oblique rule; if this modified into no longer an desire, direct rule came into strain. All kinds of exploitation were enforced with the aid of using manner of such names as McCartney, Eliot, or perhaps Admiral Perry. World records became complete of betrayal, guns, and pills in place and time, built on continental land and controlled thru marine electricity. Imperialism is the winner takes all. It actually takes time and location to carry out it.

Meanwhile, Matteo Ricci quoted Confucius, who stated, "When the world is at peace, the focal point is on people searching after themselves and does no longer comprise useless exploitation." This idea is the proper use of our vicinity and time library room to consider worldwide information. Outside the palace, a farm and town joined with telegraph traces. On the desk are spices, coffee, and lots of others. There are toy soldiers and ships on the floor. In the corner are sitting three guys assembly commercial employer leaders and politicians. They are promoting weapons and tablets, however none are to be depended on.

Chapter 18: Deception to your thoughts

We are simply in a big room with a Japanese Matsu Hito, a Turk, and a Russian attempting on one-of-a-kind garments while reputation inside the front of mirrors. Hito garments as Kaiser Wilhelm of Germany. Peter, the Great of Russian, garb as Napoleon at the same time as speaking French. Selima the Third, the Ottoman Turk, cannot determine what hat to put on. There are 3 maps at the wall. The Russian map has St Petersburg in the middle of the globe. Mahmud the Second's map has Istanbul inside the middle, and Hiro's map has Japan focused. The maps contemplated in the mirrors are backward to make matters extra complicated. When the maps are read, it's far hard to recognize French, Russian, Turkish, or Japanese translations. This situation is positioned inside the contemplated corridor of deception. Looking on the maps, Russia desires to annex components of Finland and Sweden. Japan desires to take over Korea, and

Turkey dreams as masses of Europe as possible. The maps do not display very last obstacles due to the truth those rulers usually need more land and enjoy invading special international places to get their manner.

Leaders get dressed in fancy uniforms, notwithstanding the reality that, on the identical time, they do not recognise what their dependable topics need, and feelings of resentment develop. No one desires to stay in serfdom. Factories are constructed, and metallic is produced for hundreds wars. The leaders within the Hall of Mirrors of your palace-like warfare and maps show how borders alternate. When Abraham Lincoln entered the corridor saying America may use hundreds of copper to construct a telegraph line from Washington to Moscow, Lincoln and Alexander II were assassinated. The telegraph line stopped quick on the Rocky Mountains. We are uncertain if this happened due to a scarcity of copper or if leaders were too grandiose. In time, leaders related their hotlines to speak approximately what uniforms to place on inside the next warfare.

There have been unplanned surprises. Japan destroyed Peter the Great's Russian fleet with its new strong ship 'Mikasa' built in England. The Koreans beat the Japanese with their 'Turtle Boats.' There isn't always any want to explain what came about next while fancy get dressed members within the Hall of Mirrors acquired phrase that someone had positioned the machine for nitroglycerine. All Empires constructed on expansionism and aggression came to a degree, much like the Roman Empire, in which they might really slowly fade. That is proper, mainly for the quiet human beings outside the Hall of Mirrors who determine any uniform worn is in no manner as crucial as a way to feed and contend with your human beings. Beware of political concept distortion. It can have an impact for your capacity to consider that fact.

Now we circulate directly to symbols to do not forget. These are readability rooms in the palace you are constructing.

Memory photo 1 – The Soldiers Wu[ii]

Upon getting into the Memory Palace, which direction do you skip? Is it at once beforehand? Did you look proper as you entered? If so, you can see guards (infantrymen or warriors) locked in mortal fight. Freeze this image in your thoughts, and do not permit them to move as you bypass them. After taking walks through the guards, you will see 3 greater ideograms (pix) as you increase and deepen your memory. There are inscriptions above the door, as placed in lots of Chinese houses. As you input the Reception Hall, the primary picture is Wu, that means battle. This ideogram includes meanings – the spear and to save you with the hand. The image on the proper is a warrior with an outstretched spear. Lower left is a hand protecting again and containing the arm of the thrower. The reminiscence cue within the calligraphy image Wu (warfare) consists of a opportunity for peace. We are strolling into the Hall of Wu.

You can also enter the Hall of Mirrors with many doorways even as you first enter the palace. Men try on uniforms in a unmarried nook, and in some different corner of the hall is the area for religious fervor. Here human beings get wearing a extremely good manner and are experiencing visions or imaginings they'll inform others. Magic is likewise practiced right here. Right now, numerous manifestations are going on. Air is modified into God. Earth is modified into Jesus, and water is modified into the Holy Spirit. The relaxation of advent is grow to be hellfire. One Italian priest proper right here claims heaven is filling the earth with ricotta cheese and dropping rain ravioli and marzipan. These are magical incantations. When we look for the authentic skip, we find out enough relics to build a resort. But it's far unstable to overthink or preserve in thoughts too properly on this room, for you is probably suspected of having too much magical electricity. The area is complete of subversives. They say, "Here, it is first-class to hold it simple silly." Let's go away right here earlier than the ones imagined powers come our way.

We now have a memory photo of war and peace represented thru squaddies. Ricci stood the numerous classical Roman Empire and the Turkish Empire. Catholics fought

Muslims, and the French joined the Turks to combat the Italians. Ricci looked at upgrades in war strategies and concluded the greater improvement; the more killing takes region till it can't be stopped. He attempts to go out the Hall of Mirrors, entire of unidentifiable torsos, legs, arms, and heads. Back in reception, we will journey deeper into our Memory Palace – a place containing limitless opportunities. The 2nd photo is princess Hui Hui Yao. Who is she?

Memory picture 2 – The Princes Hui Hui Yao

This image is extra complex in calligraphy as Yao has severa meanings. Yao can advocate want or need that requires pressing action because it might be completed. Ricci divided the character in half of with the useful resource of a horizontal line. The higher part interprets into the west. The decrease proper right into a lady does not explicitly mean western women. It also can endorse foreigner. Yao is a girl from the Xixia region, an historical state inside the west of China. Xixia became moreover home to the 'Hui Hui' Chinese Muslims. Xixia have become not confined to Muslims but furthermore to scattered Jews and Nestorian Christians. It have to be remembered this vicinity modified into a part of the Silk Road, where trade gadgets and cultural ideas had been exchanged. The Silk Road declined as transport through manner of sea have turn out to be economically super to the west. As using the Silk Road declined, Chinese towns decayed. Needs improved, and the Xixia tribal woman came to mean necessity. She isn't heroic, brave, or incorrect however is selfless. She does no longer rob others and is positioned in a subdued, quiet, lightly lit

Reception Room's northeast nook. The locked infantrymen are positioned in the southeast corner of the reception area. Let's permit her live. Leave Yao in her corner, separated from infantrymen to her left.

Ricci stated the Chinese used a lot gunpowder for cute fireworks that it can supply Europe with enough gunpowder to remaining severa years. Now, appearance once more on the Wu infantrymen. Is the spear advancing, or is the hand advancing or maintaining regular? If we understand the advancing spear, we are in horrible deviance. When the spear advances, a need is created as humans and belongings are destroyed.

What goes on these days? Are now not lots of human beings rioting and demonstrating because of the fact they want meals, jobs, fairness, and peace? This fact is why you cannot allow the squaddies pass. Hold your hand immoderate to stop the drones of battle and insanity. Interestingly, China isn't at war and is expanding economically and politically on the equal time as the west is locked in conflict and broke. War will no longer and can not broaden way of life till many arms save you the spear until the desires of the people are met. The tribal lady (necessity) is looking ahead for your response within the soft, quiet slight.

Necessity is likewise an instinctive reaction. Like the hand retaining decrease returned the spear, every person have this intuitive information of self-defense, which consist of at the same time as you maintain up your arm to deflect a blow. A Memory Palace is a place of intuitive notion and safety in the direction of melancholy, humiliation, and violence. The moderate slight born and held in reminiscence is intuitive, smart belief. From Yao comes the inherent, essential responsibility to fulfill the desires of these much less lucky, as informed in this proper tale of Ota Benga.

Ota Benga's darkish reminiscence and revenge

This insert is the real tale of Ota Benga to reveal melancholy exists. Ota modified right into a pygmy from Africa and is an instance of humiliation. He had no manner to protect himself and created suicide in America. There changed into no smart thought inside the way he end up treated. In 1889, the World Fair changed into held in Paris, France. At this sincere, the 'human zoo' have become created.' A 'Negro village' changed into built to show the 'civilized human beings' that sub-species lived in Africa. You might also moreover want to have a examine the African village via a stressed out fence. One Black African on display become Ota Benga, displayed with a chimpanzee as his quality associate and modified into categorised as a 'pygmy in a cage.' He traveled in his cage to St. Louis, the World Fair, and in the end, the Bronx Zoo. There he killed himself.

Now permit's turn his unhappy tale spherical. When powerful forces thieve from weaker individuals who do now not recognize ulterior motives, why need to they be established a cage? Are warring infantrymen now not examples of a double enormous? Does the Yao tribal female have any power to govern the exploitation of the awful through manner of manner of the rich? Is the arena only a playground for the wealthy? In distinctive terms, is the area turning into an area wherein the wealth of nations is extracted via a few for the few? If so, the spear movements in advance, and the slight dims across the tribal lady. Only loose hands can tear down cage doors and block spears to free Ota Benga's soul. Remember Ota's name. Humiliation works in tactics, and intuitive concept factors to the perpetrators of injustice.

Expand the Ota Benga enjoy to a everyday degree; we will find out unique thrilling times of humiliation. As the Wu sword moved in advance in Latin America, tens of tens of millions of Aztec, Maya, and Inca neighborhood people perished, growing the weight of information with it. The tribal princess requires the want to redress preceding and present injustice. The Wu infantrymen inside the meantime are Conquistadors and Samurai. The Spanish introduce racism on a worldwide, grand scale. One Conquistador said, "We will praise God and get rich." Cortez worn out the Aztecs with thirteen guns and smallpox. One hundred fifty thousand masses of silver had been stolen from Mexico and Peru and have been minted into Spanish cash, which inflated the price of the whole lot in Europe with the aid of the use of the usage of 500%.

The not unusual lifespan for a nearby silver miner declined to seven years, and people had to be replaced as they died off. Over the following one hundred years, 5 million enslaved people have been introduced to the New World from Africa. They were trapped, for that they had nowhere to transport. The colonial situation based totally totally on a caste system lasted for 300 years. Sugar changed gold and silver, and the display went on. Sugar made some human beings very fat in greater techniques than one. By 1800 liberation movements grew: 1800 riot in opposition to the Spanish, 1808 wars of independence, 1818 Chile have emerge as unbiased, and through 1824 South America became free of Spanish dominance due to the truth Napoleon invaded Spain. Simon Bolivar built Colombia, and Brazil's Empire got here into existence as slavery changed into banned. The majority had been black, and that they held their fingers towards the sword of Wu. Meeting the dreams of the horrific emerge as tough to come by means of way of as warlords, autocrats, army dictators, Santa Anna, and US Monroe Doctrine took land and sources for themselves.

One voice talking out in opposition to such abuse of energy modified into Jose Marti from liberated Cuba. He concept and stated, "In Our America, there are herbal and artificial people. Natural people are of severa shades but internal are all of the equal. The artificial people use this distinction in colour to enslave, however in fact, actually everybody are created equal". An artificial individual is much like the choose in Detroit who made a judgment no person (in particular black and terrible) has any right to loose water from Detroit, it absolutely is positioned at the shore of Lake Michigan. There is likewise the homeless lady stuck by means of manner of manner of police ingesting lunch with a plastic spoon scooped from a can of Spaghettios. The police arrested her for possession of methamphetamine. After a month, the police lab checks at the records determined simplest spaghetti residue. She have become launched. Jose Marti had stated no to poverty and racism. Understanding these mind, you will undergo in mind seeing the intense moderate at the tribal female in the nook of the Reception Room.

The mnemonic (memory) devices used in the Memory Palace is probably as an alternative simplistic magical, or naïve. They are organizational tool to useful resource in know-how forces at work international. Ricci's explorations in India describe Portugal's ownership of Goa which turned into seized in 1510 from the Muslim Sultan of Bijapur thru Alphonso de Albuquerque. The motivating forces had been the Viceroy, Crown, noblemen, merchants, and clerics who got here to run the show in Goa, which have become a base for religion, battle, and change. Hindus were subjugated. Muslims had been at the upward push, led with the aid of manner of Mogul ruler Akbar who visited Goa. He had little interest in Christianity and favored opium. How forces art work in statistics calls for a tremendous statistics of sports and dedication to the reminiscence of such occasions. Strong photographs supply up the textual content of statistics for us to conform with.

Akbar had a few detail in commonplace with the Portuguese – they every held mutual hatred for the Jews. The first Jews have been expelled from Portugal in 1497. To guide King Sebastian in his battle in competition to Morocco and avoid expulsion orders, the Jews paid bribes of 250,000 ducats to live in Portugal for ten years, but the king become killed in combat at Alcazarquiir in 1578, and the deal have become revoked. Jews needed to skip into Ghettos and had been locked in at night time. In Rome, they have been the bankers allowed to build up interest. They superior the garment exchange, production silk, folding beds, and early device gun prototypes for Rome. The Inquisition have become enforced in 1543. At Ancona, Italy, a dozen Jews had been thrown onto a pyre. The Grand inquisitor put on his black gown inside the Hall of Mirrors (consider this room?) By 1560 Jews had been fleeing to Goa and Cochin, China. Then in 1571, the Inquisition arrived in Goa. Ricci felt uneasy seeing Lutherans and 12 Jews burned to death in some unspecified time in the future and 12 greater Jews the following day. Next, the homes of the burned were confiscated and provided through using the usage of the Papacy, which had invented a profitable organization. The underlying forces of the Inquisition were greed and murder. Ricci, distraught, left Goa for Zhaoqing, China.

In China, he decided that Muslim investors have been buying and promoting on the Silk Road for years. They lived within the antique Xixia nation (princess), which the Mongols had destroyed in 1227. Muslim migrants arriving in China were Sunnis who severed contributors of the family with the Shiites in Iran. (Sounds acquainted) Muhammad's religion spread in the course of China and dotted the nation-state with many fantastic mosques in pagoda style. The Jews first arrived in China in approximately the seventh century. The Jews, like Muslims, do not consume red meat, so the Chinese known as every agencies the Hui Hui. Hui Hui is a time period for indifference closer to foreigners. The top notch European wars supposed little to the Chinese. Eventually, Ricci became time-venerated into the Emperor's Court of China. He have become given increasingly substantial court obligations till his demise in Beijing on May 11, 1601. This tale is ready Ricci's live in China.

Our Memory Palace sits on a hill bathed in smooth light, bringing us back into the Reception Room to understand the 0.33 photo. Ricci takes a Chinese name. His name in Chinese is Li which transposes to Ly and Ry, said Ri for Ricci. This Chinese image Li manner earnings. Ricci is the farmer. The farmer is located within the northwest nook diagonally across from the infantrymen locked in combat, and the Hui Hui tribal princess's left inside the northeast corner. Using his instance, we can understand the way to make the most of reminiscence benefit (Li).

Memory picture three – The Farmer Li

Ricci divides the calligraphy vertically in half of of of. We now have extra meanings. The left half of of approach grain, and the right half of method knife, blade, or sickle. Your reminiscence is growing a farmer equipped to obtain grain beside the Hui Hui and contrary the infantrymen within the Reception Room. The farmer stands organized for harvest inside the northwest nook.

Ricci is developing a few element from no longer anything. As you'll find out, that may be a exceptional approach for reminiscence education. The immobile figures suggest a time even as there has been no reminiscence or experience, as we understand the clean slate (tabula rasa). Something specific took place a long term in the past when reminiscence and belief revel in commenced out. Origins of concept got here from the primary moderate, and nothing as an explosion of existence bureaucracy passed off. Conditions were created out of a chaotic universe, and the spirit of reminiscence turned into born via necessity. Li's farmer photo shows the idea of fellowship and cooperation created to live on. This concept photo suggests caring for land, gardens, and oceans. It isn't unintentional that reminiscence has been created due to the fact it is able to produce surprising outcomes. Memory may be advanced as we are using memory to create the layout of the Memory Palace. When the way is refreshed, the reminiscence may additionally moreover spark off a multi-dimensional universe called reorganizing and restructuring in careful idea. We can construct a few issue we want without obstacles. We have started out out out from not something and might now deliver together a noble palace in a multi-dimensional universe. Let's replicate on historical memory reconstructions and notice how they're able to assist build our Memory Palace.

Li is earnings, however how is benefit measured? Is a town like Florence, Italy, extra profitable than Guangzhou, China, due to the fact one has more artwork portions? Or, are historic manuscripts held in a metropolis extra treasured than art work? What is income? How can one take advantage of seeing Michelangelo? The Catholic Church benefited from past practices. Pilgrims poured into Rome to peer grand processions of finely robed monks, surrounded by using the use of Swiss guards, taking walks from the papal palace past exemplary tapestries adorned with garlands of plant life. They were furthermore accompanied thru set up cavalry and choirs making a song on the same time as sporting gold and silver crucifixes and superb sacred vessels. Such spectacles have been alleged to daze pilgrims with suggests of wealth and pomp. It changed into like triumphal Roman warriors returning to Rome with the spoils of struggle. But terrible pilgrims may now not advantage from this at the identical time as Pope Sixtus V rebuilt Rome and completed the dome on St. Peters. Pilgrims would depart with recollections of such spectacles. Ricci ought to take into account witnessing such indicates of extravagance, which involved profits (Li) received on the fee of others (farmers). He did no longer like what he noticed but stored those mind to himself. The grand inquisitor made certain that people like Ricci held their silence.

Instead, Ricci created his Memory Palace in his head. Ricci advanced the traits placed in intuitive, mounted reminiscence and later installed them in China. This reminiscence isn't always mere invention or propaganda but is the light of popularity being passed on. The Memory Palace is positioned on the hill of facts, and what it consists of is saved for all to look time-commemorated occasions spread as they want to. This vision is the splendor of the palace – unfolding reality. Those brought approximately first-rate thru income will not realize this palace have become constructed, for they've got selected to be locked within the Hall of Mirrors, from which few escapes arise. A agency currently entered the Hall of Mirrors thru mistake and found it tough to get out. Should you stay in that corridor too extended, they'll use your head as a bowling ball to win their activity of winner takes all. Imagine if the farmer and sickle are moved from the Reception Room to the Hall of Mirrors; a few issue terrible will take region.

That is why the farmer want to live where he is and put together for the harvest. In which nook is he located?

In Ricci's worldwide, we will observe China's Li dating amongst land, knife, and profits. The Jesuits were now not mendicant to have interaction in organizations to pay their bills. The Jesuits determined worthwhile twists and turns in silk alternate that have come to be unreliable. Ricci have come to be an astute negotiator in Chinese land gives, writing that the Chinese had been now not trusted. Ricci moreover referred to that Chinese college students in Guangzhou have been interested in constructing Memory Palaces. To provoke the Chinese, he requested a duplicate of 'Mirabilia Urbis Romae,' which particular in writing and illustrations the past splendor of the eternal metropolis of Rome. This ebook become speculated to affect the royal court docket in Beijing, for it showed Europeans had a superb civilization and have been not actually mercantilists.

In Goa, Ricci created a garden of flora and fruit trees to rival Milan, Italy. In Goa, Muslims and Hindus have been prohibited from painting Christian topics, and Jesuits have been moreover limited. Under such policies, Jesuits had to discover special guide manner, and they discovered the silk alternate. They invested in trade among China and Japan. Jesuits profited from the "Nagasaki Run" in dubious dealings, which have turn out to be silk sent from China to Japan with a 25% markup. But earnings modified into not continually assured as transport gadgets through sea in leaking wooden boats is precarious. There had been many cargos out of place at sea. So an awful lot cargo changed into out of area that humans refused to art work with Jesuit silk buyers. Ricci moreover end up quick of cash and wanted some factor to sell.

Fate intervened within the shape of a clock. It changed into now not prolonged till Ricci obtained decide on with Chinese officials who desired clocks. This mechanism become the iPod of the day. Another place of interest to Ricci became land. He bought tiny homes on a riverbank. The ruler noticed the land purchases and pressured Ricci to sell them at a loss. Out of coins all another time, Ricci started making engraved sundials that have turn out to be well-known. He furthermore have emerge as a actual belongings agent for wealthy Europeans shifting to China. He fought Chinese officers to benefit gadgets and services tax exemption, and he gained. Again, Ricci ran out of cash and become saved with the resource of a pal Wang Pan who preferred a clock from Macao whilst a clockmaker from the Canary Islands arrived. Wang Pan gave him two metallurgists, and the number one European-designed clock which regulated time grow to be 'Made in China,' however it didn't artwork. Pan gave the defective timepiece once more to Ricci.

Ricci then found out that the primary spring-driven clocks had been invented, which rang at the hour and sector-hours. Some have been small sufficient to area on throughout the neck. Like their pendulum clocks, Ricci and Jesuits swung between poverty and affluence. At instances, donations and even gadgets of enslaved human beings arrived. Jesuits observed some other manner to income - human trafficking. Runaway slaves had been all over again to their previous owners for a finder's rate. All of this elements to the hassle of earnings and alternate in China. When viewing Christ crucified, a Chinese buddy of Ricci said, "it isn't always top to have someone appear to be that." Here is a cultural difference.

Chapter 19: Random perception

What are the relationships amongst Li (Ri) and the farmer equipped to attain? Ricci and the Jesuits pursued income to maintain and extend their missions in China. In Europe, traders had amassed all the first-rate wooden stands for ship advent. They did now not replant the stands of timber, and as change grew, they were forced to use green or immature wooden for deliver's planking. After ships had been launched on their voyages, the greenwood got here apart, and caulking fell out at the same time as the ocean rushed in. This manufacturing end up the principle reason the change modified into precarious, for the European farmers had reduce down the entirety in sight and did not prepare for the subsequent harvest. Blinded with the aid of the want for income, random idea made many companies fail as warring parties (infantrymen) attacked every distinctive's ships. Outside the Reception Room of the Memory Palace, the context changed.

Soldiers locked in combat keep guns; the black princess is obtainable into slavery, and the farmer has destroyed his land. They had been bringing the Reception Room into the triumphing time. What takes region? The soldiers maintain rockets, Princess Hui Hui has Ebola, and the seas are eighty% depleted of marine life. You can see how treasured readability is as positioned in reminiscence reception.

The superb issue about a Memory Palace is its flexibility, but how accurate is it? If we keep in mind the Memory Palace an area of drama, it's far going to be a combination of fact and dream. For Ricci, the fort is a intellectual vicinity, a idea for max impact in growing[iii] a a success assignment in China. If memory is a present, giving items should construct a grand palace, counting on the gadgets given. How might this utility art work? Ricci prepared present-giving to impress the Chinese Emperor with the cause of profiting by gaining want. The gift listing modified into lengthy and guarded an ostrich, carpets, mirrors, horses, clocks, wine, swords, maps, hourglasses, silver cash, art work, and one harpsichord. Eight horses and 30 companies brought those low-rate items to Beijing, however the Emperor have become incensed by way of way in their trinkets.

In pass lower back, the Jesuits had been boarded in livestock stalls. The Jesuits gave the Chinese reasonably-priced iron shields and a few sickly horses. Because of this, it is probably some time earlier than Ricci may want to hire a small house in Beijing. The Emperor knew that constructing a grand palace supposed giving extraordinary gives first. Ricci might also want to do a little component extra notable. In this example, what is truth, and what is a dream? They are intertwined. Ricci desires of an brilliant reception, however his journey outcomes in a cattle barn.

In the Memory Palace, figures in the Reception Hall appearance beforehand to greater drama to spread. If we flow the figures, the theater will certainly extend. Should armed soldiers circulate closer to the African princess, her want for food and water will boom. Should they bypass inside the route of the farmer, his flowers can be destroyed. If the princess (want) movements within the path of the farmer, no food is probably left for her. Ricci needs all figures to stay regardless of the fact that to hold a stability. He tries to create a dramatic portrait of his witnesses' social, cultural, and financial events. He knows that looking for income creates dislocation, oppression, and unintended results.

Careful recalling of sports is the reason of the Reception Room. No depend how small, any movement can be decided and recorded in memory for later do not forget. Memory sifts through dramatic factors to parent truth from dreams. The memory inside the Reception Room ought to keep its quiet balance (validity) and be saved cut free the Hall of Mirrors (drama). If there may be an interchange among the 2 thresholds, intellectual readability is misplaced as random mind seem as desires. The quit end result may be chaotic, causing faulty bear in mind. Memory calls for stillness for clarity of idea. The princess dreams the farmer to prepare the arena (mind) for the harvest of sound and crucial wondering concerning studying modern-day events. It turn out to be now not practical to speak out about contemporary occasions within the direction of Ricci's time however recalling sports drawn from the reminiscence monetary institution inside the palace have to reveal beneficial at the proper time.

Balancing your reminiscence

When figures live of their corners, it's miles a time of balanced relationships. The stability is dissatisfied via carelessly transferring spherical figures, developing chaos and resentment. Let's shipping the reminiscence Reception Room again to Europe and the Middle East to look what we will have a look at balance. This notion will serve to extend the reception region. The room is so as and however in smooth moderate, but someone from the Hall of Mirrors has moved our soldiers in advance. Nations are marching to struggle. As men are destroyed, governments promise colonies can be freed inside the occasion that they assist the conflict attempt. Many ensures are made as squaddies and horses are driven into the abyss. The battle ends, however our infantrymen have cracks, and there can be no coins to restore them. The infantrymen are despatched decrease again to the corner, however any guarantees made for their assist are forgotten. Nations are bankrupted however need to hold control of their colonies and take assets owned thru vanquished countries. The farmer has new proprietors of his land. He protests. A league is common within the Hall of Mirrors, and the farmer is knowledgeable he ought to have domestic policies. While the farmer complains, the membership tells him they'll be his trustees and could mandate future land use. Anti-sedition legal hints cowl the farmers in India. The balance is eliminated due to the fact the proper to have a fair trial is erased. When the Indian farmer tries to step forward, he's shot. It is now doing or loss of life.

The farmer Li proclaims, "They will overlook approximately me, snigger at me, fight me, after which I will win." The princess speaks from her corner, "You do not want slavery and bondage." From the hall, an Egyptian shouts, "They have seized our land and our canal. They say we are now independent as they control our foreign places affairs. Balfour is sending Zionists to Palestine. Arab land is taken and sliced up without regard for our farmers. Hertzel, Dreyfus, and

Rothschild set this up. Palestine became partitioned in 1937, and there has been no balance considering that day. Sunnis and Shiites are locked in combat in their nook even as the excellent the united states given entire independence, Saudi Arabia, is wealthy. All the relaxation are trapped inside the Hall of Mirrors and are poorer due to it." It has been a wild day in the Reception Room. Soldiers skip even as it's time to divide and rule the people. The stability is tilted and will rate the princess her strength and the farmer his conventional manner of lifestyles. Things loosen up at the same time as the squaddies skip lower returned to their nook and restore balance. If this is not achieved, hooliganism will run amok.